THE EDUCATION OF
ALICE HAMILTON

THE EDUCATION OF ALICE HAMILTON

From Fort Wayne to Harvard

—ɯ—

Matthew C. Ringenberg,
William C. Ringenberg, and
Joseph D. Brain

INDIANA UNIVERSITY PRESS

This book is a publication of

Indiana University Press
Office of Scholarly Publishing
Herman B Wells Library 350
1320 East 10th Street
Bloomington, Indiana 47405 USA

iupress.indiana.edu

Cataloging information is available from the Library of Congress.

ISBN 978-0-253-04399-3 (paperback)
ISBN 978-0-253-04400-6 (ebook)

1 2 3 4 5   23 22 21 20 19

Dedicated with gratitude to those
progressive era reformers who sought to make this
world a better place.

CONTENTS

PREFACE

ALICE HAMILTON'S PATERNAL GRANDPARENTS, ALLEN and Emerine Hamilton, helped to develop Fort Wayne, Indiana, from an early nineteenth-century frontier town to a postfrontier city. Alice grew up on the family compound on the near south side of Fort Wayne while America was experiencing its major Industrial Revolution, growing in the post–Civil War era from the world's fourth-largest to its biggest industrial power.

The Industrial Revolution came too rapidly, with the rate of change producing growing pains that caused serious harm, especially to the largely Eastern European immigrant city dwellers who supplied the labor to drive the industrial machine. The income differential between the rich and the poor widened, urban social services lagged behind population growth, political corruption increased, and big business grew more rapidly than did big government's ability to regulate it.

As a young adult, Alice Hamilton became a medical doctor and relocated from Fort Wayne to Chicago. At first she served as a physician to the poor immigrants whose transition toward being productive Americans was encouraged in Jane Addams's famous Hull House settlement (founded 1889), where Hamilton lived. While there, she gradually assumed the primary role of her life, namely as the nation's chief investigator of industrial diseases and injuries. As such she earned a place with the muckraking journalists and Presidents Theodore Roosevelt and Woodrow Wilson as a major progressive movement reformer. Such was the respect for her work during the first two decades of the twentieth century that when Harvard University wanted to add a position in the field of industrial health in 1919, they chose her as the obvious candidate despite the fact that prior to then the institution had never employed a woman professor.

Why are we three authors interested in Alice Hamilton? More specifically, why are we interested in the importance of education in explaining—professionally and personally—the woman that she had become by her full maturity? All three of us work in higher education—M. Ringenberg (Valparaiso, social work), W. Ringenberg (Taylor, history), and Brain (Harvard, public health). Two of us study aspects of education as a research specialty (W. Ringenberg, the history of American higher education; M. Ringenberg, the influence of family on educational success). All three of us took our undergraduate work in the same institution (Taylor University) in which Hamilton completed her first year of medical studies (while her father was a member of the board of trustees). W. Ringenberg, like Alice, is a native of Fort Wayne and has written two histories of Taylor University (the college operated in Fort Wayne from its beginning in 1846 until it relocated to Upland in 1893, two years after Alice Hamilton's attendance). Brain was the longtime chair of the Harvard department (environmental health) into which Hamilton's department evolved.

It was Brain who first conceived of this project. He tells how the idea of a work on Alice Hamilton grew in his mind:

> When I arrived in Upland, Indiana, in 1957 as a seventeen-year-old freshman, little did I know that my four years at Taylor University would ultimately intersect with an amazing woman, a leader of global significance, Alice Hamilton. After my Taylor graduation, I married a Taylor classmate, and we moved to Cambridge, Massachusetts, where I enrolled as a graduate student in what was then called the Division of Engineering and Applied Physics (DEAP) and has now morphed into the Paulson School of Engineering and Applied Sciences (SEAS). After getting a master's degree in applied physics, I was attracted to the intersections between engineering and physics with health and disease. Then, after a year of taking courses at Harvard Medical School, I moved to the Harvard School of Public Health and received a master's degree in radiological hygiene and then a doctor of science degree in physiology, where I specialized in lung biology. As I evolved into a physiologist studying the deposition, fate, and health effects of inhaled airborne particles, I learned that one of the heroes in the history of environmental health with a global reputation was Alice Hamilton. Even today, she is

widely regarded as the mother of occupational medicine and industrial toxicology.

Two decades ago, while continuing my career at Harvard University and as an active alumnus of Taylor University, I discovered that Taylor and Harvard were connected. While continuing my scientific career in environmental health, I eventually became chair of the Harvard T. H. Chan School of Public Health's Archives Committee and soon became fascinated with our many distinguished alumni and faculty. One of these, of course, was Alice Hamilton. Meanwhile, my college classmate and coauthor of this book, William Ringenberg, had become a professor of history at Taylor University. Bill and I sometimes met, electronically or in person. Together we gradually discovered that Alice Hamilton, like the two of us, was also affiliated with Taylor University, as were her father and grandfather.

Time passed, and both Brain and W. Ringenberg were busy with other projects, but finally, in 2015, they decided that it was time to focus on Alice Hamilton. Also at this time, W. Ringenberg asked Brain if the former's son, Matthew, could join the team as the point person for the research on the period between Alice's formative years based in Fort Wayne, 1869–97 (W. Ringenberg's focus) and her second career at Harvard, 1919–35 (Brain's focus). As Hamilton chose to pursue her medical science career at Northwestern University while simultaneously living and serving at Hull House, Chicago, the most prominent settlement house in America, Matthew's social work background (PhD, Washington University of St. Louis), career (social work faculty at Valparaiso), and geographic base (one hour from Chicago) made him a good fit to lead the study of Alice Hamilton's early career. The two Ringenbergs then became the principal researchers and writers, with Brain playing a meaningful complementary role on the project, especially in writing chapter 1 of this book.

Another major development in the history of the project was the discovery that our interests and those of the Indiana University Press could blend together nicely. Stephen Wrinn, director of the Notre Dame Press, on a Taylor visit, encouraged W. Ringenberg to contact David Hulsey of the Indiana University Press (IUP). When he did so, he was pleased to learn that the press had special interests in women's studies, public health, and prominent Indiana people and institutions.

It is important to note that this work on the "education" of Alice Hamilton uses the word *education* in the large sense, meaning much more than formal learning or even book learning. Its focus is upon the major ideas, values, and experiences that shaped her into such an extraordinary woman with such a remarkable career.

The two most significant published biographical studies of Alice Hamilton are Alice's own autobiography, *Exploring the Dangerous Trades* (Boston: Little, Brown, and Company, 1943); and Barbara Sicherman's, *Alice Hamilton: A Life in Letters* (Cambridge, MA: Harvard University Press, 1984; and Urbana: University of Illinois Press, 2003). Hamilton wrote the autobiography when she was in her early seventies after years of being encouraged to do so by her cousin (on her mother's side), Edward A. Weeks, the editor of the *Atlantic Monthly*. It is a largely objective and dispassionate study that, as its title suggests, focuses primarily upon her professional career.

Sicherman's *Alice Hamilton: A Life in Letters* is a combination primary source (131 letters written over seventy-six years—age eighteen to age ninety-six) and series of biographical sketches ("A Life in Brief") that introduce the letters and provide context for them. A careful work of scholarship, it provides details—especially about Alice's personal life—that nicely complement and enrich Hamilton's own narrative.

Our present work is a more specialized study than are the above two works, focusing, as noted earlier, upon the ideas and experiences that informed or educated the developing Alice Hamilton throughout a lifetime of learning. The most important chapter is also the longest one; chapter 6 discusses the findings that Hamilton uncovered in her industrial research and that led to major industrial reform. Her invitation to teach at Harvard (chapter 1) was the ultimate recognition of the importance of her work. Thus this book features this accomplishment at its outset.

A comprehensive list of Alice Hamilton's fifteen federal government reports, seventeen books and pamphlets, and 165 articles appears in the bibliography of Wilma Ruth Slaight, "Alice Hamilton: First Lady of Industrial Medicine" (PhD dissertation, Case Western Reserve University, 1974, 212–26) and is reproduced with permission and appreciation as the last section of this book.

ACKNOWLEDGMENTS

MANY OTHER PEOPLE CONTRIBUTED TO the making of this book, and we acknowledge them with gratitude. These include Walter Font and Richard Halquist of the Fort Wayne History Center; John Beatty of the Fort Wayne Public Library; Diana Bachman and Emily Swenson of the University of Michigan Bentley Historical Library; Valerie Harris and Adrian King of the University of Illinois Chicago Richard J. Daley Library; Steven Calco of Cornell University's Institute of Labor Relations Archives; Daniel Bowell, JoAnn Cosgrove, Lana Wilson, and Linda Lambert of the Taylor University Library; Desirae Crouch of the Taylor University History Department; Mark Robison of the Valparaiso University Library; Jeff Swisher of Hammond Gavitt High School (Indiana); Barbara J. Niss of the Arthur H. Aufses, Jr. MD Archives and Mount Sinai Records Management Program; JB Tamar Brown, Julia Greider, and Natalie Kelsey of the Radcliffe Institute, Harvard University Arthur and Elizabeth Schlesinger Library; Douglas R. Atkins of the US National Library of Medicine; and Barbara Faye Harkins of the National Institute of Health Library. Also, Sarah Wehlage served as the manuscript manager, and Jonathan Schoer (Valparaiso University, chemistry), Jennifer Bjornstad (Valparaiso University, foreign languages), and James Garringer (Taylor University, photography) served as consultants. Harvard graduate assistants Glen Krugolets and Melissa Curran aided in the acquisition of many of the photographs, especially those coming from the

Hamilton Family Papers in the Schlesinger Library on the History of Women in America. Dawn Davenport of Mount Lebanon High School (Pennsylvania) and the Valparaiso University Writing Circle critically read parts of the manuscript. Our able and readily helpful contacts at the Indiana University Press have been associate director Dave Hulsey and editor Ashley Runyon.

ALICE HAMILTON

—𝔪—

BRIEF EDUCATIONAL BIOGRAPHY

FORT WAYNE NATIVE ALICE HAMILTON (1869–1970) became a progressive era physician, scientist, and activist, investigating the health of industrial workers and crusading for improvements in their working conditions. Also, during the middle part of her career, she became the first woman faculty member at Harvard University.

Alice was born February 27, 1869, to Montgomery and Gertrude Pond Hamilton and grew up on the family compound established on the near south side of Fort Wayne, Indiana, by her paternal grandparents, Allen and Emerine Holman Hamilton. Grandfather Allen accumulated the family wealth through strategic partnerships in land sales trade with Native Americans and through other enterprises during the Fort Wayne pioneer period. This wealth funded large land holdings, including the three-square-block family compound that contained the three homes of the grandparents ("the Old House"—one of the grandest homes in northern Indiana) and their sons Montgomery and Andrew. Other uncles, aunts, and cousins lived nearby, so on any given day, as many as seventeen cousins, including Alice, could be roaming the estate grounds, playing, exploring, and discussing books. Often they found a pleasant spot to read a book from the vast family libraries, which were especially rich in classical literature and history. The parents and tutors helped with direct language instruction.

The family delayed most of the formal learning of the children until the boys were ready for college and the girls for female academy, typically in elite Eastern institutions like Princeton, Harvard, and Miss Porter's

School for Young Ladies (see table 0.1). Later, Alice and her older sister, Edith, transcended the traditional family educational terminus for daughters by enrolling in collegiate institutions—Bryn Mawr for Edith, three medical schools for Alice, and the Universities of Leipzig and Munich in Germany for both.

But just as education writ large did not begin for Alice and the other Hamilton children with their formal schooling, so also it did not end there. Alice Hamilton in particular matured greatly in her worldview and self-confidence when she began working at Hull House, when she began her investigations of the work and living environments of the Chicago turn-of-the-century immigrant community, when she accepted the teaching position at Harvard, and even in retirement when she became a public intellectual. Learning, for Alice Hamilton, was a lifelong project.[1]

Table 0.1. Chronology of Major Life Events

Birth	New York City February 27, 1869	
Childhood	Hamilton Family Estate Fort Wayne, Indiana	
Education	Homeschooling	
	Miss Porter's School, Farmington, Connecticut	1886–88
	Fort Wayne College of Medicine, Taylor University	1890–91
	University of Michigan Medical School	1892–93
	Northwestern Hospital for Women and Children, Minneapolis	1893
	New England Hospital, Roxbury, Massachusetts	1893–94
	Universities of Leipzig and Munich, Germany	1895–96
	Johns Hopkins Medical School	1896–97
	Pasteur Institute, Paris	1903
Career	Hull House, Chicago; resident physician and community and industrial health researcher	1897–1919
	Women's Medical School of Northwestern University, professor of pathology	1897–1902
	Memorial Institute for Infectious Diseases, Chicago	1902–10
	Illinois Commission of Occupational Diseases, member	1908–10
	Illinois Survey of Industrial Diseases, director	1910

(continued)

Table 0.1. (continued)

	United States Department of Labor, special investigator	1911–20
	Chicago Pathological Society, president	1911–12
	American Public Health Association, officer	1914–18
	Harvard University Medical School and Harvard School of Public Health, assistant professor of industrial medicine	1919–35
	League of Nations Health Committee, member	1924–30
	United States Department of Labor, consultant	1935–?
	National Consumers League, president	1944–49
Major Publications	*Industrial Poisons in the United States*	1925
	Industrial Toxicology, first edition	1934
	Industrial Toxicology, second edition (with Harriett Hardy) With subsequent revisions, this book became known as Hamilton and Hardy's *Industrial Toxicology*. It is now in its 6th edition (2015).	1949
	Exploring the Dangerous Trades (autobiography)	1943

General summary biographies of Alice Hamilton include Jacqueline Karnell Corn, "Alice Hamilton," in John A. Garraty and Mark C. Carnes, eds., *American National Biography* (New York: Oxford University Press, 1999), 9:910–12; Melba Porter Hay, "Alice Hamilton," in John A. Garraty and Mark C. Carnes, *Dictionary of American Biography* (New York: Charles Scribners Sons, 1988), supplement 8, 241–42; and "Alice Hamilton: Obituary," *New York Times*, September 23, 1970, 50.

ONE

—⚋—

PROLOGUE

Alice Hamilton Arrives at Harvard

WE BEGIN OUR STORY OF Alice Hamilton in the middle. She is fifty and springs to increased public attention when she becomes the first woman faculty member ever appointed at Harvard University, the oldest university in North America (founded in 1636). The *Boston Globe* headline reads, "A Woman on the Harvard Faculty!" Similarly, the *New York Times* states, "Woman in Harvard Post. Dr. Alice Hamilton First of Her Sex Elected to the Faculty." This historic event is documented by the official notification to Dr. Alice Hamilton from the President and Fellows of Harvard College, dated March 10, 1919, and signed by "Your Obedient Servant, Francis Wilson Honeywell, Secretary." This appears on a standard form with elegant calligraphy and includes blank spaces where the details of specific appointments can be inserted. Interestingly, the form includes the greeting "Sir." No one bothers to correct this salutation.[1]

This chapter tells the story of Alice Hamilton's career as a professor at Harvard from 1919 to 1935, when at age sixty-five, the traditional age of retirement then, she ended her official connection to Harvard. Her public influence continued until her death. After this chapter the book reverts to the normal order of a biography by telling the story of how she became the person and researcher that Harvard would seek to employ and then describing her career after leaving Harvard.

One could view this chapter as the beginning of a mystery. Mysteries engage the reader when they raise questions at the outset. Something unexpected happens. Why and how could it have happened? Who is responsible? The first woman faculty member at Harvard did not come from Boston or Cambridge or New York City or Philadelphia, or even Cambridge or Oxford, England. She came from a family in Fort Wayne, Indiana. She was homeschooled and then went to Miss Porter's School in Connecticut—a finishing school designed to train rich women to be "educated" wives and mothers. How did this path lead to America's oldest university?

Besides "explaining" Alice Hamilton, we must consider other components of this mystery involving Alice's siblings. Alice had three sisters, all of whom grew up in the same Hoosier environment and were subjected to the same guidance and expectations of Miss Porter's School in Farmington, Connecticut, near Hartford. Yet all four women escaped the trajectories typical of graduates of this school. None married, and all had distinguished careers of service, scholarship, and activism.

Harvard University had never employed a woman faculty member since its founding. What, then, was so unusual about Alice Hamilton that the nation's most prestigious institution of higher education would want to break that nearly three-hundred-year tradition by recruiting her to join its faculty in 1919? There is no evidence to suggest that her appointment was a result of feminism infiltrating Harvard. There were no voices suggesting equal opportunities for women, either as students or on the faculty. The entire university, including Harvard College and the growing number of prestigious graduate schools, were all-male preserves. That was unquestioned.

The rationale for Hamilton's appointment grew out of an identified need for a new component of Harvard University and the Massachusetts Institute of Technology (MIT), which began in the fall of 1913–14. This major development in Harvard's history was the creation of the Harvard University and the Massachusetts Institute of Technology School for Health Officers in the fall of 1913.[2] It arose from a carefully crafted plan developed at the highest levels of both universities. The two responsible administrators were the presidents of these two institutions: Abbott Lawrence Lowell, of Harvard University, and Richard C. MacLaurin, of the Massachusetts Institute of Technology.

This joint effort had a director, Milton J. Rosenau, but no dean. Its overall governance came from an administrative board: William T. Sedgwick, chairman, a MIT engineer and biologist; Milton J. Rosenau, a Harvard physician specializing in public health and infectious diseases; and George C. Whipple, a Harvard microbiologist. These leaders were located on both sides of the Charles River, with Rosenau's office at Harvard Medical School in Boston, Sedgwick's at MIT, and Whipple's at Pierce Hall—a building that still exists—in Cambridge.

MacLaurin and Lowell wanted their universities to have a practical impact on health and disease, especially on infection in children in Boston. They were committed to prevention, not just treatment. They realized that solutions to health problems were complex. Reducing the deaths of children required attention to the quality of drinking water, sanitation, hygiene, and nutrition. They recognized that health and longevity had more to do with environmental factors than with doctors, nurses, and medicines. As this school moved forward, it excelled because its administrators were able to carefully select the very best faculty and the most appropriate courses from throughout the two institutions. It soon offered seventy courses in such fields as preventative medicine, personal hygiene, public health administration, sanitary biology and chemistry, and communicable disease. The school's focus was multidisciplinary science with practical goals. It had a profound effect on the citizens of greater Boston and also established a model that was widely emulated throughout the United States and globally.[3]

After a few years, the school leaders realized that a major cause of disease and injury had to do with where men and women worked. Health science was beginning to understand the pathogenesis of black lung (coal workers' pneumoconiosis) as well as diseases associated with silica and asbestos. It was clear that many injuries and fatalities resulted from unsafe practices on the job. This realization led the school's officials, especially David Edsall, dean of the Medical School, to recruit Alice Hamilton. Industrial medicine and occupational health were becoming major fields of study in the early twentieth century, and the School for Health Officers wanted to develop a program in this area. They knew of no one, male or female, with industrial health research experience and knowledge comparable to that which Hamilton had acquired. So Dean Edsall, with the concurrence of President Lowell, hired Alice Hamilton, to her delight. In her humble words, "I was really about the only candidate available." Becoming the first woman faculty member at Harvard was the highlight honor of Alice Hamilton's career and one of the highlights in the history of women in American higher education.[4] Thus this book features this unique achievement at its outset. This chapter discusses the details of her initial appointment and explains her career at Harvard for the sixteen-year period when she served.

As stated earlier, Alice Hamilton's appointment did not grow out of a desire for gender inclusion. Rather, it grew from necessity. The Harvard-MIT School for Health Officers sought to include in the curriculum all factors that influenced health. Those included air, water, food, sanitation, and poverty. Early course offerings showed that where people worked was already considered an important factor as well. For example, there was growing evidence that certain industrial chemicals and particles caused lung disease.

The 1916–1917 circular of the School for Health Officers listed a course, Industrial Hygiene and Sanitation, stating that it dealt with "the various prejudicial effects of factory life on health, including occupational accident, industrial poisonings, and the effects of defective ventilation and of dusty trades." Thus, very early, the workplace was seen as a key determinant of health. However, the instructor for the course was listed as Selskar M. Gunn, who, although on the faculty at MIT, only had a bachelor of science from that same school. Moreover, he did not specialize in the field of industrial hygiene and occupational medicine. He had a broad interest in public health and sanitary science and also taught a course on sanitary law and public health administration. Thus, although there was a recognition that industrial health was a topic essential to the training of health officers, the early instructor lacked sufficient credentials and experience. This must have been awkward for Harvard. By contrast, Alice Hamilton had already spent a full decade researching occupational health and had shown that she could not only produce academic publications but also successfully advocate for appropriate legislation that saved lives. She was simply the best there was.

Dean David Edsall pursued her by inviting her to give the annual Cutter Lecture at Harvard Medical School. This move by Edsall appears to have been designed to serve a double purpose. He wanted to sell Harvard to Alice and Alice to Harvard. Bringing her to Harvard in 1918 as a Cutter lecturer was a chance to show her that many, such as Edsall and Cecil Drinker, were excited about the possibility of her coming permanently. The second, and more difficult, purpose was to showcase her and thereby convince the president and board, better known as "the Corporation," that Harvard needed her. During the previous decade, the Cutter Lecture had been presented by prominent, well-known scientists and physicians, most of whom were tenured full professors elsewhere. The selection of Alice Hamilton was a clear statement to the presidents of MIT and Harvard (and the Corporation in particular) that she was of the same caliber and was comparable in accomplishments. Her appointment as Cutter lecturer in 1918 was part of the "education" of Harvard's governing boards and made her an even more credible candidate a year later, when the school sought to make her an assistant professor.

The leadership of the School for Health Officers proposed to the corporation that Hamilton be appointed as an assistant professor. The initial response was no. Harvard not only confined its student body to the male gender; the faculty and senior administrators were also exclusively male. But Dean Edsall and other leaders at the School for Health Officers, such as Cecil Drinker, persisted, and finally, about a year later, Hamilton received her appointment. Edsall's compelling assessment was that "her studies stand out as being of unquestionably both more extensive and of finer quality than those of anyone else who has done work of this kind in the country. A very unusual and sound person."

Although trained as a physician to diagnose and treat disease, she was convinced that prevention was better than diagnosis and treatment. Her experience at Hull House and her work with the Department of Labor gave her ample background for leading this area. Her impulse to travel and collect best practices globally also contributed to her unique background. In particular, she was familiar with the prevention of work-related injury and disease in Europe, including Germany, which was highly industrialized, especially in mining and manufacturing. When her appointment began in 1919, Hamilton was fifty years old. Unusual for this time was the agreement that she could continue her field work while she also taught courses in industrial toxicology and occupational medicine.[5] Her appointment letter famously included three restrictive conditions. In her words, "I was asked to promise not to try to enter the Harvard Club which at that time did not even have a ladies entrance for wives of professors. I was not to ask for my quota of football tickets, and I was not to march in the procession at Commencement." When asked whether those conditions persisted, she said, "I suppose so. I never asked [for them] to be rescinded."[6]

Her years at Harvard and later were important because of her role as a mentor, colleague, and author. Perhaps her most famous protégé was

Harriet Louise Hardy, who became a pioneer in occupational medicine and also the first woman full professor at Harvard Medical School (sadly, Alice Hamilton never gained a promotion, retiring at age sixty-five still an assistant professor). Hardy took an interest in unusual diseases and helped to identify their pathogenesis. While studying illness at factories that produced fluorescent bulbs in Lynn, Salem, and Ipswich, Massachusetts, she discovered Salem Sarcoid, a serious, progressive, and poorly treated disease caused by the inhalation of air containing particles or fumes of beryllium. Shortly after Hardy published her first report on beryllium poisoning, Hamilton asked her to collaborate on a major scholarly work. A distinguished international student during Hamilton's Harvard tenure was Prince Mahidol Adulyadej, who became the father of public health and modern medicine in Thailand, as well as the father of the long-reigning (seventy years) king of Thailand, Bhumibol Adulyadej. Alice Hamilton also became a close colleague of brothers Cecil and Philip Drinker. Cecil was a physician and an expert in cardiopulmonary physiology as well as occupational health. Philip was an engineer who was an expert in exposure assessment. He studied the particles and gasses in the air that people breathe in various work settings. He also became famous for his invention of the iron lung, known locally in Boston as the Drinker Respirator.[7] When asked about Dr. Hamilton's Harvard career by the HSPH historian Gene Curran, Phillip Drinker said, "She was very tactical when handling management, even when they were very difficult" and also that she was "very fearless and scrupulous in the way [she] reported on [her] findings."[8]

A landmark event during Hamilton's Harvard years was her 1925 publication of *Industrial Poisons in the United States*. It discussed a novel field and became the definitive textbook on the topic. This book drew on the work she had done prior to coming to Harvard as well as on studies carried out during her academic tenure. This monograph was extraordinary in its quality and breadth. Each of its thirty-eight chapters had a bibliography and carefully crafted footnotes. It was not only a compendium of industrial poisons but also a guide to scientific investigation in the field of industrial health.

In 1934, Alice Hamilton produced the first edition of her well-known book *Industrial Toxicology*. In 1947 she asked Harriet Hardy to work with her in producing a second edition of the book; as Hardy later recalled, Hamilton told her, "My dear, I've decided that there is to be a second edition of my book, but I cannot possibly do it by myself." This book is now in its sixth edition (2015) and continues to be known as *Hamilton and Hardy's Industrial Toxicology*. It remains an important resource focusing on industrial exposures and toxic substances.[9]

Throughout her tenure at Harvard, Hamilton continued to be connected to the federal government, primarily the Department of Labor, which was originally a bureau in the Department of Commerce. Also, during this period, she was a frequent consultant for General Electric, she traveled to the USSR and to Germany, and she served on the Health Committee of the League of Nations (1924–30). She advised President Hoover and was involved with multiple groups that advocated for worker health, peace, and the poor. While a faculty member at the Harvard School of Public Health, she received honorary doctorates (ScD) from Mount Holyoke College (1926) and Smith College (1927). In addition, in 1935, Eleanor Roosevelt presented her with the Chi Omega Sorority National Achievement Award.

During her years at Harvard, 1919–35, she usually taught only one term per year. The remainder of the year, she continued to engage in a variety of industrial medicine activities, often using Hull House as her base, as she had done for the previous ten years. In her first year, Hamilton's half-time salary was $2,000. The next appointment was for $2,400, and this increased to $2,750 in 1921. Finally, on September 1, 1935, she retired and was appointed assistant professor of industrial medicine emeritus.[10]

EARLY INFORMAL EDUCATION

AS STATED IN THE PREFACE, this book is an identification and explanation of some of the major ideas and concepts that Alice learned. This chapter will focus primarily on the influence of her paternal grandparents—the family founders in Fort Wayne—and her father.

Alice learned from her grandfather, Allen Hamilton—who died five years before she was born—that wealth makes a difference. Allen Hamilton (1798–1864) was born in Northern Ireland into a family of moderate wealth and professional attainment, but when his father's illness prevented him from studying law, he came to America in 1817 (and was naturalized in 1824) to find ways to achieve financial and vocational success. Throughout his American experience he possessed the ability to link up with key partners in profitable enterprises. His insightful friend and contemporary public intellectual and public official, Hugh McCulloch, in his late-in-life memoirs, noted that from the time of Hamilton's clerical position with a Philadelphia Quaker businessman, "his career was one of uninterrupted success."

In 1823, while living in Lawrenceburg (Dearborn County), Indiana, he accompanied town founder Samuel Vance in relocating to Fort Wayne. The federal government had just opened a land office and appointed Vance as the register of land titles and Joseph Holman as receiver of public moneys. Holman also engaged in the Indian trade, and young Hamilton became the clerical assistant in the land office. Fort Wayne only recently had negotiated peace with the neighboring Native Americans, and this made inoperative its most recent and best developed fort.

Hamilton showed an ability to focus on the business ventures that at any given time were the most opportune ones. In his early career that meant the land business and the Indian trade. Especially significant was his Native American trade partnership with Cyrus Taber. Hamilton focused on both helping his clients and enriching himself. He befriended the Miami Indians in general and their chief, Jean Baptiste de Richardville, in particular. He became Richardville's business adviser and legal counselor. In turn, Richardville, who controlled the vitally strategic Maumee-Wabash portage west of Fort Wayne, became known for his keen ability to negotiate maximum benefits from the federal government for his people and himself. Hamilton, as his adviser and friend, was paid well, especially in choice land sites.

At the same time Hamilton worked with and within the white federal government system, which operated under the assumption that the land— or nearly all of it—must pass from the Native Americans to white settlers. In achieving this goal, the government highly valued Hamilton's skills in negotiating land cessions because he was widely respected and trusted by the Native Americans. In retrospect, he has been criticized by some historians for engaging in "shrewd and often fraudulent land purchases" and for becoming wealthy at the expense of the Native Americans and the small farmers. By contrast, Hugh McCulloch expressed an interpretation of Hamilton's involvement that was more typical in the nineteenth century:

> The tribes that occupied the northern part of the United States were incapable of being civilized, by which I mean, incapable of living by manual labor upon land or in shops. . . . Land is needed for grazing and cultivation. Every acre is, or will be, required for the subsistence of the human family. Territory sufficient to support a thousand Indians by hunting and fishing, would furnish homes for hundreds of thousands of industrious white men. . . . While this [gradual disappearance] is to be their fate, there is cause for national humiliation in the fact that their disappearance has been hastened by the vices, the cupidity, the injustice, the inhumanity of a people claiming to be Christians.[1]

Although the Hamilton cousins, especially Alice and Agnes, were sensitive to issues of social justice and the importance of serving the needs of the lower classes, it is difficult to find a place where they commented critically on their grandfather's role in achieving wealth through acquiring Native American lands.

As time passed and the number of Indians decreased while the number of white settlers increased, the enterprising Allen Hamilton moved beyond the earlier strategic partnerships with Samuel Vance, Cyrus Taber, and Chief Richardville to form new businesses in dry goods, milling, and banking, and new partnerships, especially with Jesse Williams and Hugh McCulloch in banking. Hamilton was president, and McCulloch, cashier and manager, of the first state bank in Fort Wayne (founded 1835), and Williams joined the two as partners of the first private bank in Fort Wayne, known as Allen Hamilton and Company (founded 1853). Williams was the chief engineer of the Wabash and Erie Canal, the key link in the waterway from the Great Lakes to the Ohio River waterway, while McCulloch developed his banking skills to such a level that he rose to the highest federal administrative office ever held by a Fort Wayne citizen: secretary of the Treasury under Presidents Lincoln, Johnson, and Arthur. Hamilton, meanwhile, had married two of his children to children of business partners, Taber and Williams, and at his death left an estate of a million dollars or more, which allowed his heirs the freedom to choose how they would live and learn.[2]

Alice learned from her grandmother Hamilton that the wealthy have social obligations no less than personal privileges. If Allen Hamilton achieved success in business, he also did well in finding a wife—both marrying into a distinguished family and marrying a person of high standards. In 1828, at age thirty, he returned to Dearborn County to marry Emerine Jane Holman (1810–89), who was twelve years his junior and from a prominent political family. Her father, Jesse L. Holman, was a lawyer, Indiana Supreme Court justice, federal district court justice, and ardent abolitionist. He was also a founder of Indiana University, Franklin College, the Indiana Historical Society, and the Indiana Bible Society; the first president of the Indiana Baptist Association and a vice president of the American Sunday School Union; and the father of William Steele Holman, who was elected a then record sixteen times to Congress and who championed an equitable land policy in the West. From her first family Emerine brought to her second family a strong sense of religious duty, social justice, and appreciation for education.

After their wedding, Allen brought Emerine to their initial Fort Wayne home in the decommissioned 1816 fort. When Emerine discovered that the young town had no organized church, she immediately moved to remedy that situation, convincing her husband to request the American Home Missionary Society to send a resident minister, preferably of the Presbyterian persuasion "in as much as they are generally better educated, and others here having predilections in favor of that order." So the Hamiltons were founders of the First Presbyterian Church.

Alice knew Grandmother Hamilton for twenty years, remembering her as "an ardent advocate of the Temperance Movement and chiefly through it of women's suffrage." She hosted in her home women's Christian temperance leader Frances Willard and suffrage leaders Susan B. Anthony and Elizabeth Cady Stanton. Before the Civil War she strongly opposed slavery, saying, "Slavery has ever been my abhorrence and everything that gives it permanence or causes its extension is most painful." Consistent with her strong antislavery views was her charitable work with the Fort Wayne African American community. She also was the founder of the Fort Wayne public library with particular concern that lower-class women have access to good literature. If Grandfather Hamilton was the family patriarch in terms of accumulating wealth and comfort, Grandmother Hamilton was the family matriarch, specializing in the love of learning and the idea that women could make the world a better place.[3]

Alice learned from the collective family group that learning itself is of paramount importance. Tom Castaldi, the official historian of Allen County, Indiana, noted of the extended Hamilton family that the trait that "tied the members together over the generations and gave them all purpose was education." The Hamiltons combined Emerine's passion for learning with Allen's financial resources to accumulate on their estate what was likely the best library in northern Indiana. Three generations of the family valued and used intensely this collection in largely self-directed programs of reading. No less important than the easy access to this collection was the passion that Emerine and others displayed toward its resources. Years later Alice recalled in her biography:

> My grandmother was a fascinating person to all of us. . . . She was a tiny person, quick and wiry, and her mind was as quick as her body. She loved reading passionately. I can remember often seeing her in the library of the Old House, crouched over the fireplace where the soft-coal fire had gone out without her knowing it, so deep in her book. Once she was perched on the top of a ladder, level with the last shelf of the

bookcase. She had meant to dust the books but had come upon one so fascinating that she lost herself in it and forgot where she was. She could enthrall us children with Scott's poems, telling us the story in prose and suddenly dropping into long passages of the poetry."[4]

Many of Emerine's children developed extensive libraries of their own, with Andrew Hamilton of the second generation and Allen Hamilton, the physician, of the third generation each possessing the largest collection in the city during their adult years.

Of her reading habits Alice noted that "most of the hours we spent indoors were spent over books," but much learning took place outdoors as well. The latter was by her mother's strong encouragement. Outdoor play and outdoor learning intersected. When the children were not reading in nature, they often were play-acting what they had been reading (e.g., Robin Hood, Knights of the Round Table, and the Siege of Troy). Or on long walks, Edith, the natural storyteller, would narrate the beginning of a classical story, then would stop at an exciting place and say, "Now you have to finish it yourself." While much of reading from the library emphasized literature and early history, many of the third generation also developed an interest in art. Younger sister Norah showed early drawing skill, and she and cousins Agnes and Jesse were to become accomplished artists. Part of Norah's training was with James McNeil Whistler.

Throughout her life Alice learned not just to read but to read—and listen—with an open mind. When during her girlhood the family spent summers vacationing at Mackinac, Michigan, one of their fellow vacationing friends was Judge Edward Osgood Brown of Chicago. Also a well-read person, Brown introduced the girls to "the literature of revolt" by such authors as playwright George Bernard Shaw and economic reformer Henry George. "Our talks with Judge Brown were 'eye openers,'" noted Alice, "in a field quite new to us, but we never thought of rejecting his ideas because they were new and upsetting. If they were true we had to accept them, and therefore they must be carefully thought out."

In a similar vein, years later in retirement Alice continued to be intellectually honest in her choice of reading materials. For example, her long list of journal subscriptions included both the communist-oriented *The Worker* and William F. Buckley's conservative *National Review*. One example of how she could modify long-held views came during the eve of

World War II, when she altered her World War I–era pacifism to support the American intervention against Hitler.[5]

Alice learned from her father that women, no less than men, should be educated well. The Hamilton fathers encouraged and provided for the best education for their daughters through private library resources and academy-level education. But some of the daughters wanted more, especially Edith and Alice. Alice's father, Montgomery (1843–1909), finally came to accept college and medical education for his daughters.

Her uncle, Andrew (1834–95), did not. For example, the latter's daughter, Agnes, wanted to be an architect, perhaps studying at Cornell, but her father vetoed the idea. If Montgomery was more progressive than Andrew in some of his ideas, he was less successful than his older brother professionally. Andrew practiced law and served as a two-term, Reconstruction-era congressman.

Montgomery had been an excellent student at Princeton and skillfully taught his daughters classical languages. Holman Hamilton (1910–80), University of Kentucky historian and son of Alice's cousin Allen Hamilton (1874–1961), thought that Montgomery was well suited to be an ancient languages professor; instead he became a wholesale grocer with additional interests in real estate and banking.[6]

LEARNING IN TRANSITION TO ADULTHOOD

ALICE LEARNED MUCH FROM TEN important women in her life: Grandmother Hamilton (see chapter 2); her mother, Gertrude Pond Hamilton (1840–1917); her older sister, Edith Hamilton (1867–1963); her first cousin Agnes Hamilton (1868–1961); academy headmistress Sarah Porter (1813–1900); lifelong family friends Clara Landsberg (1873–1966) and Katherine (Kitty) Ludington (1869–1955); and Hull House senior colleagues Jane Addams (1860–1935), Florence Kelley (1859–1932), and Julia Lathrop (1858–1932). This current chapter of the book will discuss the influence of Alice's mother, sister, and cousin plus Porter, Landsberg, and Luddington. The last three women will appear in chapter 5.

One of the most significant things that Alice learned from her mother was the importance of personal liberty. Women in particular must possess the freedom to not be restricted to domestic responsibilities. This is a remarkable emphasis from a woman who, as the oldest daughter in a family of eleven children, often had to play the role of mother, and who then, when married, gave birth to eleven children of her own. While physically busy she had the ability to be free in mind and, in turn, to not exercise a possessiveness over her children. Another major lesson Alice learned from her mother was that people should be sensitive about the needs of the larger society and not just about their own concerns. "She made us feel that whatever went wrong in our society was a personal concern for her and us."

From sister Edith, older by eighteen months, she learned that she must embrace the family reading lifestyle to be a full participant in the close-knit family life on the compound. "She was a passionate reader," noted Alice, "while I was a reluctant one, . . . she was a natural storyteller. . . . Edith and Margaret were born readers. Norah and I were not, but family pressures made us into bookworms finally; and since we saw so little of any . . . children outside our own family, the people we met in books became real to us."

Edith was her father's daughter, loving the classical languages and literature so much that she majored and taught in the area of his interest. Then, when she became a writer as a second career, she wrote with such eloquence about her ultimate passion, the ancient Greeks, that her magnum opus, *The Greek Way* (1930, 1942), assured her professional reputation as arguably the best-known classicist of her time. Alice, by contrast, was more nearly her mother's daughter, valuing her independence from the judgments of Edith, and from the family's emphasis upon language and literature. Ultimately she chose science, but in a way that allowed her to follow her mother's admonition to make a difference in the world by acting to meet human needs.

In choosing to meet human needs, cousin and best friend Agnes Hamilton also had a major influence on Alice.[1] As a harsh self-critic, Alice thought herself inferior to Edith intellectually and to Agnes in basic goodness. Agnes recruited Alice to work with her in teaching at the Nebraska Sabbath Mission School in a poor section of far west Fort Wayne. She also convinced Alice to embrace liberal political views and consider a career as a worker in the new urban Settlement House Movement. Agnes was a devotee of the writings of economist Richard Ely, who founded the Christian Social Union with its emphasis on applying Christian teachings to the severe factory and urban problems caused by the late nineteenth-century Industrial Revolution. While the two cousins were reading about the Settlement House Movement, Jane Addams, founder of Hull House, Chicago, came to Fort Wayne to lecture. After hearing her, the two cousins were even further resolved to commit to the Settlement House work, with Alice choosing Hull House and Agnes going to the more evangelical-based Light House in Philadelphia, where she continued for thirty years.[2]

Alice easily, naturally, and deeply learned to love the place called home, even though she variously defined "home" as the family compound in south Fort Wayne (1869–97); Hull House, Chicago (1897–1919 and part-time thereafter); and Hadlyme, Connecticut (1935–70 and part-time for many years before). Widely traveled and broad in interests, she was nevertheless an incorrigible homebody. Alice referred to this characteristic as her "absurd clinging to my own place," and her biographer noted that throughout her life she could feel "wretchedly homesick" when absent from her family.

Alice's Fort Wayne home was a compound more than a simple house. She lived the first few years of her life in "the big house," the stately structure of her grandparents, before moving next door to "the white house" of her nuclear family.[3] The estate also included "the red house" of her uncle and aunt, Andrew and Phoebe Hamilton, and their children. Then there was the three-square-block estate on which she, with her siblings and cousins, roamed freely and for extensive periods of time. The vast homestead lay south of Lewis Street, between Lafayette Street on the east and Calhoun Street on the west. The family enclosed the area with fences and hedgerows to separate it from the city to the north and the seemingly indefinite Hamilton land holdings to the south.

Allen Hamilton donated some of his land for civic innovations. In 1852 he gave a section south of the estate for a depot for the new Indiana and Ohio Railroad. Years later Alice's father would visit the depot to invite travelers to his home to drink sherry with him. During the Civil War, which helped to popularize the sport of baseball, Hamilton invited the use of the northwest corner of his personal estate for the new sport. In 1871 the Fort Wayne Kekiongas played what historians call the first major league game on Hamilton family–owned land south of the railroad.

One of the things which later attracted Alice to Hull House was that it was a haven, a safe place, with a family environment that served as a base from which she could go on her adventures into the city, then the state, then the nation, then to Europe. She was so comfortable there that when she accepted the offer to become the first woman faculty member at Harvard, she did so on the condition that she could spend half of the year at Hull House. Also, she recruited one of her sisters, Norah, to join her at Hull House.

Already before Alice went to Harvard at age fifty, she was planning ahead to ensure that her original family could be reunited in retirement. The Fort Wayne estate was being carved up; Edith and Margaret were living in Baltimore, where they had developed careers as educators at Bryn

Mawr preparatory school for young women; and Alice had been greatly enjoying the environment in Connecticut, where she annually attended the celebrated house parties of academy roommate Kitty Ludington. So in 1916, she asked Kitty to research and recommend an appropriate New England retirement place for her sisters and herself. The result was the Comstock Homestead in the small community of Hadlyme, twelve miles from Kitty's home in Old Lyme and immediately adjacent to a ferryboat terminal on the Connecticut River. By the time Alice retired from Harvard, she had already spent eighteen summers at Comstock. Margaret and Norah joined her there. Edith also did for a while but then became attached to her own summer home in Sea Wall, Maine, near the scenic Acadia National Park. Longtime family friend Clara Landsberg moved in with the sisters as a part of the Hadlyme household. She had been a close friend of Margaret at Bryn Mawr College and an eighteen-year roommate of Alice at Hull House. Alice learned to experience the depths of friendship with her. Then, in 1920, Alice's aunt Phoebe Hamilton and her daughters Agnes, Jessie, and Katherine, accepted Alice's invitation to relocate close by in Deep River, Connecticut, a few miles from Hadlyme. Like the Hadlyme homestead, the Deep River home was near the Connecticut River, between Hartford and Long Island Sound. The children of male cousins frequently visited. Thus, once again Alice had her family together.[4]

If Alice's century of life had produced three major places called home, it also featured four secondary homes. These were Veraestau, near Aurora, Dearborn County, Indiana, the girlhood home of Grandmother Hamilton; Mackinac Island, Michigan, the summer vacation home of the Andrew Holman Hamilton and Montgomery Hamilton families (two cottages), which Alice enjoyed beginning when she was age ten; the campus of Miss Porter's School for Young Women, where for two years Alice developed close friendships along with intellectual interests; and the stately 227 Beacon Street (Boston) home of Ernest Amory (an innovative surgeon and medical reformer) and Katherine (social activist) Codman, with whom Alice lived during her Harvard years and where she developed a close sister-like relationship with Katy.

Veraestau (shorthand Latin for spring, summer, and fall) was the fondly remembered childhood home of Grandmother Hamilton, to which she returned whenever she could. She viewed the estate, situated high above the Ohio River, as an especially healthy environment. Grandson Allen Hamilton recalled that "she thought the climate had wonderful health properties and when anyone was ill she would bundle them in a carriage and take them to Veraestau." Grandfather Hamilton bought and enlarged the home in 1839, and seventy-five years later his daughter Margaret enlarged it again. Since 2004, Indiana Landmarks has owned the property.

Of her girlhood summer home, Mackinac Island, Alice wrote, "We came to love it passionately." As with the Fort Wayne estate, she came to associate it with outdoor reading, particularly poetry reading with "pine cliffs . . . , the moonrise over Lake Huron . . . , deep, deep woods; sweep of sky . . . , and water lapping on a pebbly beach."

Ten of the Hamilton young women through two generations followed the family tradition of attending Miss Porter's School in Farmington, Connecticut, which served as something of a Yale University for ladies. Graduates of the school over the years included young women with names like Vanderbilt, Pulitzer, Rockefeller, Bouvier, Biddle, Bush, Forbes, Kellogg, and Pillsbury. The graduates held "love and loyalty" toward the school, Alice stated in her autobiography. Years later the school reciprocated that sentiment by naming a residence hall "in honor of the strong-minded and strong-willed group of Hamilton sisters and cousins . . . appropriate role models for all on campus." When Alice enrolled, the school's founder, Sarah Porter, was in her seventies and toward the end of her years as leader of the school. Sarah's brother, Noah, had a similarly long career at Yale, where he served as professor of moral philosophy and then as president. Alice elected to enroll in the mental and moral philosophy courses, which used Noah Porter's textbooks. Such advanced courses at most four-year colleges were reserved for the seniors. Alice's favorite professor was classics scholar Nathan Perkins Seymour, the grandfather of future Yale president, diplomatic historian Charles Seymour. The influence of Sarah Porter on Alice was less from formal instruction than by example of personality—integrity, self-control, and clear thinking. Then there was the marvelous environment that the school provided for meeting and enjoying and learning from a broader range of girls than Alice had ever known. One new friend in particular stood out, namely Kitty Ludington. At the end of their first semester of school, Kitty asked Alice and Agnes to room with her. Agnes thought

Kitty "the nicest girl in the school." At the end of that semester, she recorded in her diary the emotional trauma of leaving school and separating from those with whom one had developed close bonds: "This is the most mournful day of all the year for the girls that are not coming back. When I rose up from prayers, I found Kitty on her bed—in tears." As an adult, Kitty hosted many famous house parties in her Lyme, Connecticut, Ludington House Mansion. These began in 1912, with Alice and Harvard Law School professor Felix Frankfurter as charter members. Alice attended many of these parties, and they introduced her to many Eastern intellectuals (e.g., Walter Lippman, Herbert Croly, and Charles and Mary Beard), broadening her thinking and paving the way for her later career as a public intellectual.

Alice greatly enjoyed living with the cordial and childless Codmans in Boston. She found Ernest both stimulating and lovable, but with Katy she developed an endearing friendship: "It is quite impossible . . . to tell how much she did for my life in Boston . . . , giving me a precious comradeship. . . . We went to meetings of all kinds, . . . foreign policy luncheons, statehouse hearings on everything from Birth Control (my specialty) to the Abolition of the Death Penalty, which was one of hers." The Codmans loved outdoor life. While Earnest was a serious hunter, the three of them regularly traveled thirteen miles to their Ponkapoag cabin for Sunday cookouts and hiking. When Felix and Marion Frankfurter invited her to accompany them to hear Roland Hayes, the legendary tenor, she wrote to friend, Clara Landsberg, that "it was worth sacrificing one of my Ponkapoag Sundays, and few things are."[5]

Alice learned that as an adult she could have a rich family life, broadly defined, without the limitations that she thought marriage would place on her desire for adventure, travel, and a meaningful career. The single most significant factor in Alice's decision to not marry was the influence of her mother.[6] Alice Pond Hamilton did not directly counsel her daughters to not marry, but she did teach them, as noted earlier, that "personal liberty was the most precious thing in life." Furthermore, Alice's parents taught the girls through their wide reading that there was a large exciting world beyond the family compound. The family library filled the girls' minds with dreams of faraway places and compelling adventures. Also, the girls heard their parents talk about their own extensive experiences living in Europe, where they had met and where their first daughter was born. Later

Alice was to travel to Europe herself, early and often. The significance Alice placed on these travels is suggested by the importance that she gave to them in her autobiography—with six chapters devoted to five of her European trips. It is significant that when, as very young women, Alice and Edith learned that family financial misfortune meant that they could no longer plan to live on inherited wealth, the obvious solution to them was not marriage but career.

Beyond the family influence, society in general in her day said that women could not have both marriage and a career, at least not simultaneously. So Alice chose for a career and adventure and travel. Her choice appeared quite natural and without trauma, although sometimes imbued with a sense of guilt for not helping her family at home as much as did some of the other Hamilton women. More difficult to understand than Alice's decision to not marry is why nearly all of the women in Alice's generation of the family also remained single. Of the twelve Hamilton women cousins, only one—the youngest—married.[7]

In the area of religion, the youthful Alice learned to embrace her mother's emphasis on the Episcopalian ritual more than her father's emphasis on Presbyterian theology. From both parents she accepted the centrality of Christian ethics (e.g., honesty, fairness, discipline, and unselfishness). The mature Alice, like sister Edith, ultimately came to identify with eastern Quaker-like beliefs and practices.

The family faithfully attended the First Presbyterian Church, which was located a few blocks north of their home, and they carefully observed the Sabbath. Alice enjoyed the change-of-pace day and the opportunity to mix with people outside of the family compound. "The Bible was more familiar to us than any other book," Alice noted later in her autobiography. "It is so deeply imprinted in my memory that more often than not when I hear it read in church I can keep just a little ahead of the minister.... Quite apart from one's faith or lack of faith in the teachings of the Bible it seems to me that an education which leaves it out is thin and poor."

Alice responded less positively to another part of her religious education, namely the harsher points of Calvinism. Her father loved to study theology and made Edith and Alice learn the Westminster Catechism, which Alice referred to as a "heathenish production." Edith also recoiled against the Westminster Creed, which she compared unfavorably with the ideas of Socrates and Plato. Jesus was also a hero to Edith, but not so many of his interpreters. By contrast to their father's religious teaching, Alice recalled, "We learned Psalms for my mother and the Sermon on the Mount and the first chapter of St. John." Then in the summers in northern Michigan, the family attended a small Episcopal church favored by her mother, Gertrude. The girls liked the latter church more than the Fort Wayne Presbyterian one. It also was in an Episcopal church in Dresden, Germany, that the parents had married.

When Alice did her advanced medical study at Michigan, her demanding schedule caused a decline in her religious discipline, but not in her religious instincts. For example, she expressed objection to the religious cynicism and even disparagement of her major professor, George Dock, even while admiring his professional skills.

Alice had a different temperament than that of her older sister Edith, and this influenced their different approaches to religion. Alice wrote to Edith's longtime close friend and biographer, Doris Fielding Reid, that "Edith had intense emotions which sometimes puzzled and distressed my matter-of-fact disposition." Edith could be up—as when describing with passion and eloquence her beloved Greeks, and down—as when experiencing a deep depression. By instinct, Edith was a philosopher; Alice, a social worker. Edith solved religious problems in her mind; Alice, with religious zeal, tackled immediate practical problems in the rough and tumble of life, inspired by her faith in "the supreme importance and worth of every soul in the sight of God."[8] In their reactions against Calvinism, neither sister embraced the common Evangelicalism of the nineteenth century, with its free-will-theology corrective to determinism, although their cousin Agnes did.[8]

Alice did retain aspects of Calvinism-like determinism. When in 1895 Edith underwent dangerous abdominal surgery, Alice wrote to Agnes that she neither prayed nor afterwards gave thanks because "it seemed so firmly fixed in my mind that all had been planned . . . long ago . . . for the best and wisest, that whether for death or for life God was good." In another letter to Agnes in 1919, she declared, "I cannot believe in the effect of prayer except on the person who prays. . . . The whole Quaker theory appeals to me, the absence of fixed dogma in matters of faith and the insistence on strict dogma in matters of conduct, the emphasis on individual liberty, the reverence for the inward light of others as well as one's own, the repudiation of authority. . . . Maybe I ought to openly join people with whose religious views I am in sympathy."[9]

Alice learned of the major involvement of her grandparents and parents in the development of Fort Wayne educational institutions, both public and private, even though they frequently desired to provide better educational opportunities for their own children, both through home schooling in their earlier years, and advanced learning in elite, usually out-of-state institutions.

Grandfather Hamilton helped to recruit the city's first teachers, served as the city school board president, and was a trustee of Methodist-affiliated Fort Wayne Female College (which became Fort Wayne College in 1855 and Taylor University in 1890). Allen and Emerine sent their oldest son, Andrew Holman Hamilton (1834–95) to the nearest Presbyterian-sympathetic college, Wabash, and their second son, Montgomery (1843–1909) to the most prestigious Presbyterian college, Princeton. Both sons also attended Harvard University and studied in Germany. The three daughters, Mary (1845–1922), Ellen (1852–1922), and Margaret (1854–1911) attended Miss Porter's School, also Presbyterian sympathetic in nature. Third-generation family members attended colleges of similar Congregationalist/Presbyterian orientation, such as Williams in Massachusetts and Yale, plus Harvard.[10]

After Allen and Emerine died, the family divided part of the family compound to enable the Fort Wayne public schools in the early twentieth century to build a new high school structure next to the Old Mansion. The city had operated a high school on another site since 1867. In 1922 the city renamed the school as Central High School when it added a new high school, South Side. Five years later it began North Side High School. In the 1930s when Central High School needed more space, an expansion to the west came at the expense of demolishing the old mansion.

Like his father, Montgomery served on the trustee board of the Methodist college ("Old ME" was its affectionate name) at the west end of the city. By the time of Montgomery's tenure, the institution's name had changed to Taylor University, and it had merged with the Fort Wayne College of Medicine. This early-1890s period is the era when the student body included both Taylor's most famous student ever, Samuel "Sammy" Morris of Liberia, Africa, as well as Montgomery's daughter, Alice, who studied for one year in the "Fort Wayne College of Medicine of Taylor University," and who may be the Taylor student to later emerge with the greatest impact on American public policy. In the original Sammy Morris film *Angel in Ebony* (1954), one of the most dramatic scenes involved President Reade, Vice President Christian Steman, and trustee Montgomery Hamilton huddled in a room discussing institutional problems when in walked Sammy Morris, fresh from New York City, whose presence offered spiritual encouragement.[11]

MEDICAL SCHOOLS

ONE OF ALICE HAMILTON'S MOST daunting and decisive realizations came in her later teen years, as she gradually processed the idea that her father's financial reverses, through the dissolution of his business partnership with Alexander C. Huestis, meant that in her developing adulthood she no longer could presume to live comfortably from the family wealth. Rather, she would have to acquire a vocation. She chose medicine.

Alexander C. Huestis (1819–95) was the first and fourth president (1847–48, 1850–52) of Fort Wayne Female College (after 1855 Fort Wayne College; after 1890 Taylor University). He was also a professor of mathematics and natural philosophy, 1847–52. Following his tenure with the college, he served a thirty-three-year career as a Fort Wayne wholesale grocer, the last twenty of which were in partnership with the younger Montgomery Hamilton. After the failure of the business in 1885, both men devoted themselves to their respective literary efforts. Huestis, who in 1849 had published *Principles in Natural Philosophy: Mathematically Illustrated and Practically Applied,* just before his death completed a manuscript for a complete Shakespeare concordance.

The 1885 failure of the Huestis and Hamilton firm had immediate and long-term effects on Alice Hamilton and her family. They could no longer spend freely. Father Montgomery had to protect the remaining financial resources. He could still support the education of a determined and purposeful child, but he was careful. Furthermore, the children would have to prepare for careers that would allow them to be self-supporting.

When sisters Edith and Alice returned to Fort Wayne from their two-year programs at Miss Porter's School in 1886 and 1888, respectively, they made their vocational decisions. Edith's choice to study and teach the classics was natural, while Alice's decision to pursue the medical field was not. All of her life Alice had studied language, literature, and history. Science she had avoided. Now she had to play catch-up.[1]

When Alice Hamilton realized that she wanted to become a doctor, she learned that she would have to overcome multiple barriers to achieving her vocational goal: her lack of prior training in the sciences, her father's concerns about her seriousness, and the reduced family finances for affording her desired first-class medical training. To remove her basic deficiencies, Alice, living at home, studied chemistry and physics with a high school teacher and pursued biology on her own. Still needing to prove herself to her father, she enrolled in the local medical college, with which he was affiliated as a trustee. Her study in the Fort Wayne College of Medicine came in the first year, 1890–91, after the medical college had merged with the liberal arts college, Fort Wayne College, to become Taylor University.[2]

Fort Wayne College of Medicine (FWCM) had begun in 1879 as a split from the Medical College of Fort Wayne (MCFW) when the latter was reluctant to embrace such reform ideas as a two-year curriculum—increased from one year—and the use of dissection with the traditional lectures. Of the twenty-four Indiana medical schools chartered in the century after 1806, FWCM was one of the best and longest lasting, continuing until the age of medical school consolidation in the early twentieth century, and eventually becoming part of the Indiana University School of Medicine. When Hamilton began what was to become her single year of study at FWCM—now "the Medical Department of Taylor University"—it was the only medical school in northern Indiana. It offered a three-year curriculum (to become four years by the turn of the century) and a teaching faculty of fourteen physicians, it practiced gender equality in admissions, and it published a respected medical journal.[3]

The medical school offered its lectures in the main Taylor building while the students conducted their applied laboratory and clinical work at the city hospital, the nearby St. Joseph Hospital—which was the largest general hospital in Indiana, and the medical school's free clinic. By the late nineteenth century, the city possessed seven railroad lines, and the medical college students as well as the professors of surgery served the victims of railway accidents. Alice's father, Montgomery, served on

the trustee board of both the medical college and the parent organization, Taylor University, and two-decade dean and professor of surgery
of the Medical School, Christian B. Stemen, had just become the acting president of the university. While Hamilton mostly studied anatomy
(including dissection), a major component of the first-year curriculum,
one professor allowed her to assist in his operations, and another, Dr.
George W. McCaskey, who later served as president of the Indiana State
Medical Association, had her examine patients with him. The year after
Alice Hamilton's attendance, the Taylor medical school purchased the
large former home of Hugh McCulloch on Superior Street to refurbish
for its new headquarters. The strength of FWMC, noted an historian of
Indiana medicine, lay with its clinical faculty, including Doctors Herman
A. Duemling, Albert E. Bulson, George W. McCaskey, Charles R. Dryer,
and Miles F. Porter, who "were widely recognized as among the best men
of the state." The last three of these were on the faculty when Alice Hamilton was a student.[4]

    When considering where she might enroll for further medical studies,
Alice discovered that there were few medical schools in the country that
were both open to women and possessing of the elite academic level that
her family had come to expect.

    Following Hamilton's year of private tutoring in science and a succes-
sive (and successful) year of study in the local Fort Wayne Medical College of Taylor University, her father was persuaded that she was seriously
resolved to become a doctor. One of her biographers has him telling her,
"Alice, I am now convinced that you really want to study medicine. You
must have the best medical education open to women." And between 1892
and 1897, she succeeded in gaining entrance to, and excelling in, two of
the most highly regarded medical colleges in the country, Michigan and
Johns Hopkins.

    By the 1890s, women had been studying medicine for over a genera-
tion—sometimes in women's medical colleges, sometimes in the regional
proprietary medical colleges, and sometimes in the elite coeducational
institutions. Even when admitted to coeducational schools, they often
faced the prejudices of the male-dominated Victorian society. For example, Alfred Stille, the president of the American Medical Association
in 1871 and the College of Physicians in 1883, thought that women were
"unfitted by nature to become physicians because of their 'uncertainty of

rational judgement, capriciousness of sentiment, fickleness of purpose and indecision of action.'" When the University of Michigan became co-educational in 1870, the medical faculty were resistant, but this posture had changed by the time of Hamilton's enrollment. Nevertheless, Alice Hamilton, herself at Michigan, talked of the "half-veiled ridicule on the part of men that one feels all the time, even while acknowledging that there was no overt 'sex antagonism.'" While the Johns Hopkins Medical School admitted women from its opening in 1893, three years before her year of postgraduate study, it was only because a critical last-minute major gift required that the institution be coeducational.

When Alice, with her sister Edith, sought to study for a year in Germany in 1895–96, she found the process of admission and study for her as a woman "a long and difficult" one at Leipzig and Munich, not negotiable at Berlin, and congenial only at Frankfurt. Overall she found the year disappointing for multiple reasons. For example, she was not certain that she had learned much medical science that she did not already know.

By contrast with her Germany experience, she found Michigan and Johns Hopkins professionally fulfilling, even exhilarating. Michigan and Johns Hopkins, together with Harvard and Pennsylvania, were at the forefront of the major transformation of American medical education between 1890 and 1910. Prior to this period of reform, the typical medical school operated with open admissions, few students with college degrees, no written examinations, one year of study (often repeated during a second year), minimal hands-on learning, and few full-time professors. The reform movement was led by doctors who had studied in Germany and embraced the German emphasis on a scientific approach to education. The changes that resulted included (1) entrance requirements including a bachelor's degree; (2) professors who were full-time scholars including researchers (as opposed to practitioners who taught on the side); (3) an expanded and graded curriculum of four nine-month years; (4) more specialized subdivisions of study beyond medicine, surgery, and obstetrics to include gynecology, pediatrics, dermatology, genito-urinary diseases, laryngology, ophthalmology, otology, psychiatry, and hygiene; (5) written in addition to oral examinations; and (6) much greater emphasis on laboratory and clinical work.

At Michigan, German-trained scholars who led the scientific reformation included dean and biochemist Victor Vaughn, pharmacologist John J. Abel, physiologist William Howell, and bacteriologist Frederick Novy—all of whom were Hamilton's teachers. In addition, she worked closely with clinical medicine specialist George Dock, who led the Michigan effort to greatly enhance the clinical and laboratory components of the curriculum. It was Dock who convinced her to balance her Michigan experience and long-range plans with a year as a hospital intern before pursuing a career as a nonpracticing medical scientist. Accordingly, she spent most of 1893–94 at the New England Hospital for Women and Children.

On the return of the Hamilton sisters from their year of study in Germany, Edith had a job offer, but Alice did not. Edith became the headmaster of the Bryn Mawr Academy in Baltimore. With no job prospect and Edith living in Baltimore, Alice decided to continue to prepare herself vocationally by enrolling in the Johns Hopkins Medical School, also in Baltimore and widely considered the leading medical college in the country. Her primary area of study was pathological anatomy with Simon Flexner, although she also enjoyed learning in the relaxed laboratory environment from the celebrated William Osler, a major pioneer promoter of clinical medicine. Flexner was the younger brother of Abraham Flexner, who, despite not being a medical man, wrote the most famous American report on medical practice, the Flexner Report of 1910. His report, *Medical Education in the United States and Canada*, both endorsed and further promoted the transformation of medical education into a uniform, scientifically based system. Alice Hamilton would have enjoyed earning a PhD, likely in bacteriology, from Johns Hopkins; however, the university did not grant doctorates to women. What the year at Johns Hopkins did give her was professional support to teach pathology at the Women's Medical School of Northwestern University.[5]

So Alice Hamilton was unusually well trained to practice and teach medicine, and she had a professorship waiting for her. But the primary reason she went to Chicago in 1897 was to live and work as a resident volunteer at Hull House, which long had been her dream. The Northwestern position provided her income but not fulfillment. In her words, "teaching and carrying on research would never satisfy me. I must make for myself a life full of human interest." Hamilton wanted a life of adventure, and Hull House provided this for the rest of her career. It was her year-round home for twenty-two years (1897–1919) and then her part-time home for sixteen years thereafter, while she was teaching one semester per year at Harvard University.[6]

The American Settlement House Movement that Alice found so compelling was one of the most significant responses to the many major social problems that developed as the United States was becoming the most industrialized nation in the world. In particular, the movement arose to serve the large-scale Southern and Eastern European immigrant population that flooded to American urban areas to meet the rising demand for factory workers. The growth of business surpassed the ability of the small nineteenth-century governmental structures to adequately regulate it, and the growth of the population in the major inner cities outdistanced the ability of government to provide sufficient social services such as housing, water, sewage, transportation, education, health, and protection.

It was into this void that there emerged many private social service agencies, mostly Protestant, to serve the mostly Catholic immigrants. In place already before the third wave of immigration (1880s–1915) were the Evangelical YMCA and YWCA, the rescue missions, and the Salvation Army centers. These were followed by the Social Gospel movement, led by Walter Rauschenbush, and the Institutional Church Movement with its urban outreach programs. The Catholics, inspired by James Cardinal Gibbons and Father John Ryan—and later Dorothy Day—developed their own versions of these services. On a smaller scale, so did the Jewish leaders. Thus the Settlement Houses had many partners in serving the inner-city poor.

Just as the YMCA and the Salvation Army spread to America from England, so also the Settlement House Movement had as its prototype the East London Toynbee Hall, which was founded in 1884 by an Anglican clergyman, Samuel Barnett, and his wife, Henrietta, and was named for the Oxford historian and reformer Arnold Toynbee. After American ethicist Stanton Coit worked at Toynbee Hall for several months in 1886, he opened the first American settlement house, Neighborhood Guild, in the Lower East Side of New York City. Similarly, Jane Addams and Ellen Gates Starr visited Toynbee Hall in 1888 and returned to America to open Hull House a year later as the third settlement house in the United States. The movement spread rapidly in America, with the six houses of 1891 growing to over one hundred by 1900 and to almost five hundred by 1920. Hull House became the most recognized of all of these. The progressive era was the peak period of the Settlement House Movement. It eventually merged with the neighborhood and community center movement, and after a period of decline it was revitalized and redefined by the Lyndon Johnson–era War on Poverty.

The typical settlement house resident worker was a young, idealistic, single woman from an affluent family and in possession of a skill. Frequently she was a teacher, social worker, vocational counselor, or health worker. They sought to provide "scientific philanthropy" (more than direct relief), assisting the impoverished to develop their skills, employment opportunities, cultural appreciation, and quality of life. Accordingly, the settlement houses offered classes in English, home economics, and arts and literature for adults; day care nursing, kindergartens, playgrounds, and youth clubs for children; healthcare clinics; and meeting halls for ethnic clubs and trade unions. They also were active participants in the Progressive Movement efforts to achieve governmental regulations to improve the lot of inner-city workers and residents.[7]

This, then, was the settlement house environment in which Alice Hamilton was to grow and mature in her learning and service during her career, beginning in 1897.

FIVE

LEARNING SELF-CONFIDENCE
AT HULL HOUSE

THROUGHOUT MUCH OF ALICE HAMILTON'S early adulthood, she was plagued with self-doubt about her intellect, effort, and virtue. These doubts coalesced around her need to find a meaningful venue to make a contribution to the world. During her time at Hull House and particularly as she saw increasing opportunities in investigative toxicology, she discovered such a venue. In discovering a venue, she discovered and cultivated confidence in herself.

Alice Hamilton remained a harsh self-critic throughout her life. Although slow to report achievements, she did not avoid mentioning them entirely. It is noteworthy that the ratio of self-criticism to self-recognition gradually shifted in favor of the latter during her tenure at Hull House. To treat this improving self-image as a simple continuum of doubt-confidence in which she moved toward confidence would be an oversimplification, however. Multiple factors beyond self-doubt and realistic recognition of accomplishments were at play. The most prominent of these factors included a sense of noblesse oblige, as instilled in her youth; the idea that women could make a meaningful difference in improving the lives of others; and a tendency to compare herself primarily to people who best embodied her ideals.

Considering the era in which she lived and the attitudes of many men regarding women in positions of authority, it might be tempting to assume that the barriers she experienced as a woman should be included on the above list. But that appears not to have been the case. While at the University of Michigan, she compared Ann Arbor favorably to Cambridge (Massachusetts) because students (male) did not stare at her in

Ann Arbor, demonstrating that she certainly noticed her treatment as a woman. Reflecting decades later on her time in Germany, where the attitudes were generally more negative toward women, Hamilton described her time there as mostly pleasant but "sometimes exasperating for an American [woman]. I had to learn to accept the thinly veiled contempt of my many teachers and fellow students." She noted various examples of being pushed off the sidewalk, forcibly moved at the opera, and reminded that women were needed in subservient roles. While she occasionally complained, these experiences do not seem to have impacted her self-image, confidence, or determination. One possible explanation is that she had simply too many examples of influential women successfully navigating social barriers to believe it could not be done.[1]

Alice Hamilton's upbringing, particularly the influence of key women, certainly shaped her expectations of herself. As mentioned earlier, both her paternal grandmother, Emerline, and to a lesser extent, her mother, Gertrude, each motivated by deep religious convictions, instilled in Alice a sense of responsibility for helping others. Emerline's influence was twofold. Emerline's own generosity extended to individuals in need who made personal requests, communities such as the African Americans of Fort Wayne, and national causes such as temperance and suffrage. Moreover, Emerline personally knew many of the most prominent national suffragettes and hosted them when they were in Fort Wayne. Thus, in the process of teaching Alice noblesse oblige, Emerline simultaneously demonstrated that women were quite capable of influencing individuals, communities, and national policy.[2]

Exposure to powerful and influential women continued and assumed a more personal form at Hull House. In the spring of 1895, Alice, along with her sister Norah and cousin Agnes, attended a speech by Jane Addams, already the most famous settlement house leader in the United States. The presentation, which took place at the Fort Wayne Methodist Church, was likely a version of Addams's oft-used "The Subjective Necessity for Social Settlements," in which she outlined the two miseries—the obvious misery of the urban poor and the hidden misery of wealthy, educated young people, primarily women, who were forced, by convention, into trivial lives. Social settlements were an opportunity to redress both these miseries. Although "miserable" was not precisely how Hamilton characterized her life, she already struggled with finding an appropriate outlet for her mind, energy, and commitment. Two years after her Fort Wayne speech, Addams granted her a room at Hull House. If Hamilton hoped to quickly immerse herself in cutting-edge social reform and learn from leaders in that arena, she was not disappointed. Within hours of her arrival, John Peter Altgeld, former governor of Illinois and the pardoner of the surviving prisoners associated with the Haymarket Riots, spoke to a large crowd and engaged in a lively discussion with Addams, Florence Kelley, and Julia Lathrop, the three women from whom Alice Hamilton, gratefully, "learned much."[3]

While each of the three women was a source of inspiration and education, they were also sources of intimidation (presumably unintentional) and a repeated reminder to Hamilton that she was unprepared to "be of use," at least in the manner and to the degree to which she aspired. Each already possessed confidence and a sense of accomplishment; Alice Hamilton did not. Even before Alice Hamilton's arrival, Florence Kelley had already translated the work of Friedrich Engels into English, systematically researched child labor conditions in Chicago, and played a major role in passing legislation outlawing child labor and limiting women's work days to eight hours. In her autobiography many years later, Hamilton described the commitment Kelley inspired: "It was impossible for the most sluggish to be with her and not catch fire." This observation corresponds in perspective to her letter to Agnes in March 1898, which described Kelley's defining qualities as "bigness and manliness and warm-heartedness." By contrast, later that same year, Alice characterized herself using imagery of a frightened child desperately holding on just to feel safe.[4]

In some ways, Julia Lathrop's example was more compelling and subsequently more personally difficult for Hamilton. Julia Lathrop, in contrast to Kelley, hated conflict and maintained a calm, "disinterested" approach to social reform. In short, her personality closely resembled that of Alice. However, as Hamilton described the situation in those early years, that hatred of conflict resulted in "cowardice" for Hamilton, but "never with her [Lathrop]." Lathrop's advice, which Alice Hamilton would later embrace as her own strategy, was that "harmony and peaceful relations with one's adversary were not in themselves of value, only if they went with a steady pushing of what one was trying to achieve." Not long after her arrival at Hull House, Hamilton accompanied Lathrop to an insane asylum. The supervisor there was openly hostile. By tactfully listening and respecting the challenges he was facing, Lathrop was able to "soften him." Hamilton thought to herself that she would have been quite satisfied with this accomplishment, but not so Julia Lathrop. She then proceeded to gently but clearly confront him about the shortcomings of his current administration. To Hamilton's surprise, he listened, with "startled meekness."[5]

Even fifty years after first encountering Jane Addams, Alice, in her autobiography, continued to express a substantial degree of reverence for her. She begins with an admission of being unable to provide an adequate description: "I cannot describe Jane Addams . . . I seem to only get bits of her." Nevertheless, Hamilton makes an attempt to summarize those "bits." She spoke of intellectual integrity, pragmatism, and a disinterest in personal accolades, particularly if they might interfere with service to others. Alice Hamilton went on to delineate "two conflicting traits" in Jane Addams that she believed made her a great woman while simultaneously causing her great pain, namely, her felt need for comradeship and harmony and an integrity that prevented her from agreeing with the majority in many crises. Having lived and worked closely with Addams both at Hull House and on peace missions to Europe, Alice Hamilton saw how deeply criticism and conflict hurt Addams, but also how they never altered her determination to follow her convictions. One can observe this dualism of Addams in major public social events, such as the Haymarket Riot or World War I, where she positioned herself for the workers and for a just peace. But perhaps the best illustration was a more personal one. Alice Hamilton described how a young couple had treated Jane Addams with contempt, born of an unnamed fear. Not long afterward, as this couple was grieving a personal loss, Addams invited them to Hull House in an effort to console them. Hamilton was indignant and reminded Addams of the earlier insults. Addams's reply was simply, "Why yes, that is true, they did. Strange isn't it." There was no righteous indignation, no self-pity. Nor did she minimize or excuse their actions. She simply chose to act charitably.[6]

Alice Hamilton's tendency to compare herself almost entirely to people who were idyllic in some quality inevitably led to self-deprecation. With Florence Kelley the ideals were her fire and enthusiasm, and many accomplishments. With Julia Lathrop it was her tact and unswerving determination to do the uncomfortable in order to achieve the good. With

Jane Addams, international reputation aside, it was nearly everything: her self-sacrifice, unswerving commitment to advocate for the rights of all people, ability to inspire others to their best efforts, and seemingly limitless compassion.

Hamilton's pattern of comparing herself unfavorably to people who embodied an ideal did not begin in 1897 when she arrived at Hull House. During her youth and young adulthood, she idealized her older sister Edith and her cousin Agnes. As Sicherman explained, "Her older sister Edith was intellectually precocious and her cousin Agnes almost oppressively good." Even at age twenty-three, in a letter to Agnes, the impact of her unfavorable comparisons was evident. "Your last letter smote my conscience, as yours generally do. Edith's letters always make me feel how utterly ambitionless I am." Edith's brilliance, particularly in languages, was obvious as early as Alice's first memories. At a young age she had memorized Macaulay's *Lays of Ancient Rome*, six multistanza poems, and the Westminster Catechism. At age seven she began reading Latin, and shortly thereafter, ancient Greek. At Bryn Mawr University, Edith earned the school's highest recognition, the Mary E. Garrett Fellowship, which paid for her studies abroad with Alice.

By 1897, when Alice Hamilton took rooms at Hull House, Edith had already found a professional expression for her brilliance. Alice and Edith had traveled together in the mid-1890s to study their respective fields at the universities of Leipzig and Munich. During that trip, while Alice was still completing her medical education and dissatisfied with her own lack of focus, Edith received an invitation to become the headmistress of Bryn Mawr (high school), the only private women's secondary school in the United States that intentionally prepared all its graduates for university studies.[7]

Cousin Agnes, meanwhile, had dutifully returned home to Fort Wayne, where she immersed herself in various ministries, most revolving around the Nebraska neighborhood mission and the Fort Wayne branch of the Young People's Society of Christian Endeavor. The latter was an interdenominational group that encouraged wholehearted commitment to Christ and Christian living, including service to others. By the end of 1894, she ran "Noon Rest," a program for working women that served tea and lunch. Agnes held administrative roles with the "League of Clubs," the Temperance Society, and the Ladies Society. She was also a key founder of the YWCA of Fort Wayne. She was eventually elevated to president of the Nebraska mission, a post she held for three years. Part of her responsibilities included recruiting more privileged members of various congregations to serve in these parachurch organizations, an interesting parallel to the Settlement House work to which she would eventually devote her career. In her correspondence with Agnes, Alice refers to Agnes successfully recruiting six volunteers and convincing their Presbyterian church in Fort Wayne to purchase a building to support the Nebraska mission.[8]

Growing up, Agnes and Alice had comprised two-thirds of the "Three As" (their cousin Allen Williams being the third), a recognized subunit of the Hamilton clan, united by common affection, mutual support, and the fact that all were born within a five-month period (Allen, October 11, 1868; Agnes, November 21, 1868; Alice, February 27, 1869). Not surprisingly, they served as points of comparison for one another as well, particularly between the girls. For both Alice and Agnes the comparisons nearly always favored the other.

Alice viewed Agnes as nearly ideal ethically: "I think that, take you altogether you are the finest girl that this wicked world contains just now and, if I am not better for having known you, it is my own fault." What exactly that ideal meant can be inferred from Alice Hamilton's other letters. While interning at Northwestern Hospital for Women and Children in Minneapolis, Alice provided detailed descriptions for Agnes of patients, professionals, and procedures as well as her own reactions to each. In one letter, Alice responded to Agnes's description of Alice's work as a mission of healing with a self-deprecating description of how she (Alice) actually experienced her work. Alice didn't "think of it as a mission of healing at all," but rather felt like "the backers of the victorious man in a prize-fight feel." Agnes' tendency to interpret activities through religious ideas only made Alice feel worse because her own experience of the same events was that of a bystander whose ego or wallet was on the line. In a letter to cousin Jessie, Alice expressed confidence that Agnes would sacrifice herself for the good of the family, an issue that reemerged several times and on which Alice Hamilton vacillated between admiration and criticism, probably reflecting her own ambiguity about her decision to energetically pursue a career instead of a more domestic role.[9]

Agnes, in turn, saw herself as Alice's (and Allen's) intellectual inferior. Fifteen-year-old Agnes recorded in her diary, "My lessons are so hard this year that I never get one moment to myself except on Saturday and yet Alice's are just as hard and even harder (she is much further advanced than I am) and she don't have to study near so much. I wonder if I am so much stupider." Shortly thereafter, in response to her sense of inadequacy, Agnes imposed a plan of discipline on herself, which may have contributed to Alice's evolving belief that Agnes was her moral superior. She would rise each day at 7:00 a.m., study before school, spend her lunch hour alone with her Latin books, and then spend another hour and a half of "serious reading" after school. To allow herself the energy to maintain this rigorous schedule, she religiously maintained an early bedtime. Whether this regimen produced tangible results for Agnes, it appears not to have changed her view of herself vis-à-vis her cousins. "I feel so childish and I know so very, very, little, I am really quite alarmed when I hear Allen and Alice talking about men and deeds that I have never heard of or that I merely know by name."[10]

Clearly Alice Hamilton held very high expectations for herself. It is a near certainty that those expectations drove her to persist in her education in an era that discouraged university education for women and in which even her enlightened father needed substantial convincing and evidence to support her. Those expectations almost certainly were a factor in her willingness to seriously consider uncomfortable philosophical, political, and religious ideas, and they likely drove her to try new tactics of persuasion that she found thoroughly unpleasant. However, when those expectations were not met, she could be very self-denigrating. "Inexperienced" and "ignorant" were among her comparatively mild self-accusations. In a letter to Agnes, she was more absolute, saying that she "did not influence a single person (while at Hull House)." This was in stark contrast to the impression left on numerous children who followed her down the street while calling out, "Doctor Hammel." [11]

As difficult as these self-doubts were for Hamilton, it would be a mistake to assume that her early years at Hull House were largely unpleasant. Counterbalancing her intellectual and personal challenges were a growing sense of connection to her new family; the satisfaction of being a part of something important; and participation in activities that were a pleasant diversion, namely theater, music, and exercise.

Exercise in particular was a valuable therapeutic component in Alice Hamilton's life in Chicago. She thoroughly enjoyed cycling, as evidenced by the not infrequent references to it in letters to Agnes. "I think I could not live through these days [of heat in the city] if it were not for the thought of a bicycle ride in the evening.... Last Tuesday I went for the first time on mine and have been every evening since. It is simply delightful." Sundays nearly always included a bike ride. Also at Hull House, one of her earliest responsibilities was as director of the Fencing and Athletic Club, a post she held from at least March 1898 through May 1899. Although little is known about her view of this role, the fact that she maintained it over a year, while rotating through other responsibilities much more quickly, implies it was of some interest. Evidence of Alice's belief that exercise was essential to mental health comes from a later letter to Agnes referencing what sounds like Norah's recovery from a severe psychotic episode. Alice reported, "Many changes, I am almost sure, are for the better ... She took swinging, rapid, walks."

Exercise therefore contributed to Alice Hamilton's increasing sense of well-being and confidence during her time at Hull House. Three other factors that played a significant role in this transformation were finding a vocation that met her demanding standards, gaining experience in leadership and organizing roles, and developing her skills for collecting evidence to use when presenting cases. The first factor was explained earlier (see chapter 4), while a discussion of the latter two factors follows now.

During her early years in Chicago, Hull House offered Hamilton a wide variety of avenues to lead and serve. She assumed the leadership of the fencing club; an organization called Anatomy for Artists; the Sunday-evening lecture series; and, along with police officer George Murphy and Jessie Binford of Hull House, the cocaine investigation. She founded the Alice Hamilton Club for teenage girls, designed around reading and game playing, a hygiene and nursing program, and a well-baby clinic. Furthermore, she encouraged residents to creatively generate opportunities themselves. She later parleyed her experience and training into a leading role on the Committee on Midwives, sponsored jointly by Hull House and the Chicago Medical Society.

Alice Hamilton's insistence on evidence as the basis for decision-making pervaded all aspects of her work at Hull House. In administering her well-baby clinic, she showed an adaptability to her clients and their circumstances based on what she saw to be effective, even if it contradicted her medical training. That training, for example, taught her that infants, prior to teething, should only consume breast milk. By contrast, the Italian mothers who visited the clinic fed their babies many foods, including bacon, bananas, and eggs. In spite of her initial misgivings, she eventually acknowledged that the babies suffered no ill effects, and therefore she shifted her focus elsewhere, such as to sanitary issues.[12]

Hamilton's relationship with self-confidence was complex. As shown in this chapter, she clearly experienced many moments of doubt and self-deprecation. She measured herself against the most intelligent (Edith Hamilton), religiously devout (Agnes Hamilton), inspiring (Jane Addams), and politically daring (Florence Kelley) people she knew. Perhaps even more difficult, she compared herself to Julia Lathrop, a slightly older woman with a similar upbringing and disposition who nevertheless had found a purpose, presence, and confidence that the younger Alice Hamilton largely lacked. Yet she would not even have made the comparisons without an awareness of her talent and potential. Her frustration with herself only made sense because she knew that she was capable of making a substantial contribution to society. Hull House gave her role models, but it also gave her opportunities—opportunities to try her talents, to fail, to adjust, and to see herself succeed. Her experiences at Hull House made possible her later achievements as a scientist, social scientist, reformer, and educator.

SIX

—ɷ—

INVESTIGATING THE
DANGEROUS TRADES

ALICE HAMILTON SPENT TWENTY-TWO YEARS as a full-time resident of Hull House. As explained in chapter 5, she transitioned from a very
self-critical new professional who often felt rudderless to a more objective,
confident, and purposeful investigator with a national and international
reputation. Key to this transformation were the research skills that Hamilton intentionally developed and honed during the Hull House years.
Although she applied these skills in her various political, economic, and
social critiques, they were most obvious in her scientific investigations of
industrial toxins—the investigations that led to her becoming the first
woman to teach at Harvard University. These research skills also shaped
her evolving view of human nature and human society.

Two specific investigations were particularly formative for Hamilton: the typhus epidemic and the trafficking of cocaine to youth. In each
she applied methodology from her medical training as well as the interpersonal skills she'd begun learning in her youth, both of which she developed further at Hull House. In each investigation she learned valuable
lessons from her successes, and even more so from her embarrassing and
frustrating failures.

Alice Hamilton returned from the family vacation home in Mackinaw
Island, Michigan, in the summer of 1902 to one of Chicago's worst typhoid
epidemics. Typhoid regularly afflicted Chicago in that era; for example, it
had killed 212 Chicagoans in July, August, and September 1901. That was
not an uncommon number. But in 1902, during those same months, the

disease killed 402. Moreover, the spike in deaths occurred almost entirely in the Nineteenth Ward, where Hull House was located. She was both curious and concerned, as were many of the Hull House residents. Jane Addams, convinced that Alice Hamilton was the resident best qualified to investigate, encouraged her to research and report on the cause.[1]

To better understand the regular outbreaks of typhus as well as the sharp rise in deaths in 1902, an explanation of the water and sewage systems of turn-of-the-century Chicago is necessary. Five pumping stations, located from two to four miles into Lake Michigan, supplied all the water to Chicago. Sewage from the city was pumped directly into the lake, near the shore. If the lake was calm, little or no sewage reached the pumping stations. However, heavy rainfall typically pushed sewage out farther into the lake, and in those cases it frequently reached the pumping stations. The water was tested daily at each station. Each pumping station was scored each day as either "good" or "bad" based on contamination levels.

Dr. A. R. Reynolds, Chicago's commissioner of health, compiled the official report explaining the increase in deaths. He theorized that an extended dry spell had resulted in a large amount of sewage settling near the shore, only to be pushed out to the pumping stations when the rains finally arrived. He explained that the October 1901 through February 1902 period was unusually dry, but the March 1902 through June 1902 months were "the wettest season on record." Thus eight months' worth of sewage was floating around and into the pumping stations by midsummer.

Hamilton included an excerpt of the Reynolds report in her own report. She accepted his evidence and analysis as valid. She then identified why his ideas were inadequate, which led to the explanation of her own theory. Alice Hamilton would use this technique of co-opting and then moving beyond a recognized expert many times in the future.

It is noteworthy that both Reynolds and, to a lesser degree, Hamilton showed a casual acceptance of a level of contamination entirely unacceptable by current standards. During the crisis, only 44 percent of the pumping station days were deemed good, meaning over half the water piped to the city was contaminated. Yet the normal rate of 72 percent, used as a point of comparison, while clearly better, was also problematic. Over one-quarter of the drinking water in Chicago was contaminated with dangerous levels of sewage under normal conditions, a rate entirely unacceptable in twenty-first-century thinking. Alice Hamilton's report did, however, mention the inadequacy of the information provided by the daily measurement system. This caused enough embarrassment that the Chicago Health Department felt the need to challenge her assertion on a technicality, which she publicly rebutted, pointing out both the misleading nature of their statement and that their assertion ignored the larger problem that residents were not being given enough information to protect themselves.

In the next section of her report, Hamilton argued why Reynold's theory was inadequate to fully explain the extra deaths that summer. Her most convincing piece of evidence was that the Nineteenth Ward, which comprised one thirty-sixth of the city's population, accounted for one-sixth to one-seventh of the cases during the typhoid crisis. The adjacent Eighteenth, Twentieth, and Fifth Wards were also hit especially hard. Clearly something else was happening, something specific to that area. At this point she employed a technique she would often use in the future: putting forth possible explanations and then arguing why they were not valid. These included personal hygiene, overcrowding, the specific pumping station supplying that part of the city (it also supplied portions of the city barely affected), and infected milk (but the milk distributors also supplied low-typhus regions).

After dismissing these other explanations, Hamilton then explained her own theory: typhus grows in standing sewage water and is spread to humans largely by flies. She provided a general description, including how the Nineteenth Ward was worse than other areas in several regards. Key problems were an older sewer system that easily overflowed and private yards that were commonly lower than the streets, meaning that as the water settled, a residue of excrement was left on and around people's homes. Alice Hamilton then introduced her methods in a way that was aimed at a medical audience, yet accessible to the general educated reader. She described five sewage disposal methods. These were, in descending order of safety and cleanliness:

1. Regulation compliant, modern plumbing (48% of the houses in the Nineteenth Ward)
2. Modern, yet flawed or broken, plumbing (7%)
3. Outdoor water closets with rain water runoff to flush them into the sewer (22%)

4. Privies with sewer connections but no regular sources of water to "flush" them (11%)
5. Privies with no sewer connection (i.e., outhouses) (12%)

With such a variety of sewage systems, one might expect houses with modern plumbing to be safer. Yet in the Nineteenth Ward, household plumbing and typhus deaths were unrelated. Instead, deaths were associated with the plumbing throughout one's immediate neighborhood. This clue pointed to a mobile culprit, something that could spread disease easily within a few-block radius. She knew of one that had been identified as the transmitter of typhoid to over twenty thousand soldiers in the recent Spanish-American War: the housefly. However, a theory alone was not sufficient. So with the help of Hull House residents, she explored the toilet facilities in the Nineteenth Ward, this time collecting flies. Flies from a variety of settings (homes/yards with standing sewage, ones with dried sewage, indoors, outdoors) were collected and tested with varied methods (e.g., bouillon, gelatin-agar). Many of the samples gathered from the flies tested positive for typhus. Others were inconclusive, which may reflect the technology of the time.[2]

In summary, Alice Hamilton employed methods that would be common in her career. She familiarized herself with the most recent scientific thought on the subject; conscientiously searched for multiple possible explanations; ruled out those that could be dismissed logically; and then immersed herself in the data gathering process to test the remaining ideas, regardless of personal hazard. She was intentional about transparency regarding her methods and results. She sought results that hinted at a practical solution to improve the quality of life (in later studies she would be more explicit that the point of the research was to improve and save lives). Her work had clear ethical implications, but the language she chose was not inflammatory or damning of either the residents or the authorities. While the last point may be, to some extent, a result of publishing in a medical journal, it also shows a shift from her earlier letters, in which she is quicker to criticize other people's moral failings. As Hamilton immersed herself in the study of industrial poisons, her sympathies with workers vis-à-vis employers grew. Yet her style in confronting employers increasingly showed the influence of Julia Lathrop: being direct and unapologetic without needless moralizing. In all these respects her methods for investigating typhus were nearly identical to those in her later work.

Yet there was a critical difference between her typhoid report and her later studies. Alice Hamilton's typhus research was arguably her most celebrated single study. Her ideas rightly led to the dismissal of negligent city inspectors, including a new chief inspector, and to an improvement in sanitation services. In spite of this, she was ashamed of the study. Soon after the health department reorganization, evidence was found, but concealed by city officials, that a sewer line had broken and was leaking directly into the water supply for the Nineteenth Ward. Although the sewer line break was never proven conclusively to be the cause of the increased typhus, Hamilton and the few people who knew of the sewer line break assumed that it was the primary cause of the increase in deaths. It appears that her conclusion about flies was not altogether wrong, but rather insufficient to fully explain the additional deaths. Alice Hamilton's oversight changed how she reported from that point forward. In all future research she took an exceptionally cautious approach, only drawing conclusions when the evidence was irrefutable.

The cocaine investigation, two years later, began with a distressed mother requesting help for her addicted thirteen-year-old son. Once Hamilton began her investigation, it was clear that the scope of the problem was much broader than one struggling teenager. Within a few years the investigation and the subsequent efforts to combat the illicit sale of cocaine would involve police officers, judges, physicians, police courts, municipal courts, the *Chicago Daily Tribune*, the Chicago Bureau of Charities, the Illinois Bureau of Justice, the Illinois State Board of Pharmacy, the Illinois Pharmaceutical Society, the Chicago Medical Society, the Illinois House of Correction, the Illinois State's Attorney, the Illinois General Assembly, and Governor Charles Deneen.[3]

Illinois's first effort to control the sale and distribution of drugs was the Pharmacy Act of 1881. It contained three key components. A list of controlled substances was created (e.g., opium) that could not be distributed to anyone under fifteen years of age. Pharmacists were further responsible for only dispensing "for legitimate purposes." Any infractions could be punished with a $5 fine. One year before the Hamilton investigation, the legislature passed a second Pharmacy Act (that of 1903). This act added cocaine to the controlled substances list, and customers could only receive prescriptions without refills. Fines were increased to at least $50, and no more than $200, for the first offense; and to $200 to $1,000 for any subsequent offenses.[4]

On July 8, 1904, the previously mentioned mother arrived at Hull House, pleading for help for her thirteen-year-old son. She explained that he was rapidly losing his health due to his regular use of cocaine. The residents agreed that her request merited an investigation. Jessie Binford, who had recently begun her sixty-year tenure at Hull House and would later administer the Juvenile Protective Association, assumed the lead in the investigation, with Julia Lathrop and Alice Hamilton also playing critical roles. Both Binford and Hamilton credited Officer George Murray with being of considerable help. Murray focused on obtaining evidence to convict the sellers rather than prosecute the victims. The investigative team began by befriending the boys in the area and gaining their trust. Binford recorded that they found addicted boys throughout the Hull House neighborhood, at John Worthy School, and in the local jail. They established relationships with twelve or fifteen (depending on the source of the report) of the addicted boys, all of whom were teenagers. Through the boys and discreet observation, the team learned the pattern of sale and distribution. To generate "customers," druggists gave away free cocaine to curious boys (and occasional girls) until a habit was formed. Druggists then began charging the youth, occasionally making the sales directly but more often using paid intermediaries; sometimes those intermediaries operated in the store, but more often on the street or even from their homes.[5]

By autumn of 1904 the investigation team had made surprising progress. Although one druggist was clearly guilty, only his clerk had been convicted (and fined $50), and the druggist could not be held legally responsible for the clerk. More promisingly, a second druggist was arrested and agreed to testify against several others, and those others were subsequently brought to trial. On October 15, 1904, Hull House sponsored a conference for the purposes of ending the sale of cocaine to minors and providing reasonable care for those already affected. Judges; doctors; representatives of the Bureau of Charities, Bureau of Justice, and State Board of Pharmacy; and other professionals concerned about children's issues participated. They were joined by four mothers of addicted sons. At the time of the conference, most of the twelve addicted boys associated with Hull House had been treated in some way. Several were sent out of town to be in healthier environments. Another six received three weeks of treatment at Presbyterian Hospital, followed by a month in the country.

Those who had been addicted only a short time made rapid progress. The prognosis for the longer users was, according to the conference report, "most discouraging."[6]

Following the early arrests, some of the druggists changed strategies. The 1903 law outlawed the sale of cocaine to (1) youth under fifteen and (2) adults except for "legitimate purposes." However, alpha-eucaine and beta-eucaine, synthetic drugs that caused the same psychoactive effects, were not on the controlled substances list, and were thus entirely legal. Once in court, even dealers who had probably sold cocaine were being released because their lawyers demanded that the prosecution show evidence that they sold cocaine, not eucaine. Hamilton's medical and scientific background was particularly useful in responding to this challenge. Miss Binford and Officer Murray confiscated boxes of drugs from the youth, and Hamilton tested the confiscated property. There was not a definitive chemical test at that time, but she was able to prove a substance was cocaine if it caused the pupils of a rabbit to dilate, which eucaine did not. The defense attorneys began swaying juries with accusations of animal cruelty, so Alice Hamilton began testing it on herself, causing at least one colleague at her lab to worry that she had developed a tumor.

In April 1906, Jessie Binford provided a follow-up report that was largely optimistic. Testimony from three boys had led to thirteen convictions. The Illinois Board of Pharmacy lawyer had even begun actively prosecuting wayward members. Specific examples were provided of a druggist who was fined $1,000 (presumably for a second offense), warrants for two "colored" men who served as intermediaries, the arrest of a saloonkeeper on Randolph Street based on a *Chicago Daily Tribune* investigation, and twelve other pending cases also based on *Tribune*-discovered evidence.[7]

By the beginning of 1907, the mood had changed significantly. For reasons not fully explained, city ordinances, not the state law, were often the basis for prosecution. Fifteen cases had been successfully prosecuted. Yet when a new city ordinance that superseded the earlier one was passed, all fifteen cases were dismissed. Furthermore, the new ordinance was so unsatisfactory as to be useless, and many subsequent cases were also lost. Only "a few" convictions, including one of the "Druggist Dahlberg," were successful, according to the *Tribune*. However, the continued efforts at prosecution and the negative publicity, while they did not stop the cocaine trade, at least forced it into the shadows. Undercover reporters disguised as customers had revealed an elaborate system of passwords and false businesses (e.g., some posed as real estate agents) used to continue to sell the harmful drugs.

An indication of Alice Hamilton's rising stature was a *Tribune* story in which "Dr. Alice" is quoted along with the Illinois states attorney and a judge as authorities on the need for a new, stricter state drug law. By November 16, 1907, proposed legislation was near passage, "embodying the suggestions of Dr. Alice Hamilton." That legislation became the Pharmacy Act of 1908, which added all cocaine derivatives and synthetic substances resembling cocaine to the controlled substances list, required all pharmacists to keep records on file for five years, prohibited the distribution of controlled substances to anyone who was addicted, allowed fines of up to $1,000 for a first offense or three months to a year in jail, and revoked the license of any pharmacist convicted.

Years later, Alice Hamilton described how, after the new law was passed, they won thirteen cases. However, a "clever lawyer" found a loophole and exploited it to have all thirteen overturned. She reported, with fatigued satisfaction, that eventually an effective law was passed (presumably the Pharmacy Act of 1915) that closed the remaining loopholes, due largely to the efforts of a Catholic priest who also served as the director of "the Bridewell," a temporary holding prison.

The cocaine investigation contained several lessons for Hamilton. Perhaps most significantly, she grew even more suspicious of the justice handed down by the legal system. The unquestioning trust she had placed in authorities as a young woman was replaced with a tension. Bureaucratic structures were needed, and working with them was necessary. At the same time, she came to expect those structures to fail more often than not. A second lesson was closely linked: the value of patience. The law did eventually work, and the education and self-governance of pharmacists also improved dramatically. The third lesson was an additional step in self-confidence. A wider audience respected and valued her integrity and ability in her role as investigator. It was a role with which she was finally growing comfortable.[8]

Although not obvious at the time, 1907 was a critical year of transition for Alice Hamilton. The cocaine investigation required less of her time, and her attention shifted to a new issue, industrial disease. Sicherman estimates that it was the year Hamilton read Sir Thomas Oliver's *The Dangerous Trades*, a book that so shaped her thinking that decades later it would form the basis of the title of her autobiography, *Exploring the Dangerous Trades*. Also significant in her growing interest in industrial diseases were the numerous stories she heard from neighborhood visitors to Hull House, including stories of CO_2 poisoning of mill workers and palsy among painters who, contrary to the claims of prominent physicians, knew the cause was lead in the paint.

Alice Hamilton also followed closely the story of several workmen dropped off at a Chicago pumping station in Lake Michigan. A fire had broken out in the station and, because the rescue efforts did not reach them in time, the men were left with the tragic choice of burning at the station or freezing to death in the lake. The immediate tragedy aside, what occupied her thoughts was how the families of the men were treated. The press and much of the public lauded the company for generously paying all funeral expenses of the deceased men. But the widows and children, who were typically destitute thereafter, received no support or compensation from either the company or the state or federal government. In contrast, widows of industrial accidents in Germany were cared for by the government to a degree that allowed them to raise their children and seek other sources of income to avoid poverty.[9]

Throughout 1907 and 1908, Alice Hamilton immersed herself in the topic of worker's benefits and protection generally and industrial disease in particular. Her reading quickly showed her not only that nearly all the research on the topic was coming from Europe but that workers' rights and safety in the United States lagged significantly behind those of Japan and most of Europe. In some nations meaningful regulations had already been in place for a generation. At the time Hamilton was familiarizing herself with this literature, the United States' involvement consisted only of fatality lists in various heavy industries and isolated newspaper reports, particularly those of more gruesome deaths. Only one United States systematic research study, the Pittsburgh Survey, had even included industrial safety (but not industrial disease) at all. The Pittsburgh Survey, modeled on earlier, largely sociological surveys of US cities by Jane Addams and W. E. B. Dubois and Charles Booth's London Survey, was the most comprehensive such study of a US city to date. It examined the sociological, economic, and safety issues in the world's largest steel-producing city, and one of its primary authors, John Commons, would play a meaningful role in drawing Alice Hamilton into this field.[10]

A detailed comparison between the United States and other nations in 1908 is difficult because so little information on industrial poisons had been gathered in the United States at that point. The one exception was the use of white phosphorus (also called yellow phosphorus) in the manufacture of matches. The phosphorus report was not released until 1910, while Hamilton was researching lead in Illinois. However, it was authored by another key professional contact for her, John Andrews, who visited Hull House and shared his report with her while he was still gathering data. By 1909 six nations had outlawed the use of white phosphorus in match manufacturing, and Great Britain had legislated in 1910 to do the same. Finland and Denmark had outlawed it in the early 1870s. In addition, eight other countries regulated it thoroughly, requiring precautions such as separate rooms or air-tight containers for different parts of the manufacturing process, designated wash and lunch areas, protective clothing, medical doctors as factory inspectors, ventilation rules, and prohibitions against child labor. Of those with regulations, Italy, Belgium, and Austria-Hungary were debating whether to prohibit white phosphorus entirely. Spain, which had no regulations at the time, was also considering prohibition. Even Russia, not known for its concern for worker safety, taxed white phosphorus heavily enough that most manufacturers had switched to the insoluble and safe red phosphorus. Japan, one of the nations that regulated the manufacture of white phosphorus matches, exported them worldwide, especially to China and India. Swedish and Norwegian officials (who had strict safety rules) had explicitly stated that they were willing to outlaw white phosphorous but did not do so because they feared competition from Japan. By comparison, the United States had no federal legislation. At the state level, bills had been proposed in Wisconsin and New York. Both failed to pass.

Phosphorus necrosis of the jaw, colloquially named phossy jaw, was the most visible manifestation of phosphorus poisoning and was common among workers in early factories where matches were manufactured. It resulted from the toxic effects of white phosphorus accumulating in the mouth, especially along the mandible and maxilla. The process is thought to be the result of white phosphorus reacting with water; carbon dioxide; and amino acids, such as lysine, to form amino bisphosphonates that then lead to the ailment. The toxic effects of phosphorus caused progressive bone deterioration and prevented the body from repairing the damage. Early symptoms include toothache and gum swelling. As the process continues, the jaw steadily disintegrates, leading to the disfigurement associated with advanced phossy jaw. Eventually, the tissue necrosis, the spreading of infection beyond the affected area, and brain damage lead to death. The experience is extremely painful. Those suffering with phossy jaw were historically left with a horrible choice: surgical removal of part or all of the jawbone, or death.[11]

The first public evidence that Alice Hamilton was closely monitoring the developments in industrial toxicology appeared in May 1906 in Charities and the Commons. Hamilton's article on tuberculosis among laborers was one of several short pieces with a broader article on the association between industrial occupations and infectious diseases. Her article appears to bridge her soon-to-be-concluded career as a bacteriologist and her not-yet-discovered career as an investigator of industrial disease. In this article she highlighted the relationship between poverty and increased risk for disease. Specific predictors of tuberculosis included unsanitary working conditions; low wages; fatigue; and long, irregular work hours. She showed some familiarity with British and German reports of occupations (e.g., stonecutting) and environmental factors (e.g., poor ventilation) associated with premature death. However, her descriptions of the manufacturing processes are lacking in detail, and her recommendations rather timid.

Hamilton's next publication related to industrial diseases was titled "Industrial Diseases with Special Reference to the Trades in Which Women are Employed." It did not appear until September 1908. The scope of this article was much broader and the tone more authoritative than her 1906 writing. She demonstrated confidence while discussing safeguards in various European nations and how rates of disease had dropped and why. By contrast, she noted both a lack of research and concern in the United States. She addressed lead, mercury, arsenic, phosphorus, and rubber processing. Hamilton then identified which industries were particularly dangerous for women vis-à-vis men. Considering how cautious Alice Hamilton had become about making assertions without overwhelming evidence, it is clear that she knew her subject thoroughly. A commonly expressed sentiment among industrialists about workers was that if the work was too dangerous, they should just quit. She concluded her article with a scathing critique of that idea. She stated that blaming workers was analogous to a sea captain telling his sailors, "If you don't like the ship, get overboard."[12] Writing this article, probably more than any other single action, brought her to the attention of key social reformers who were just beginning to notice the social problem of industrial poisons.

In particular, she was noticed by Charles Henderson, professor of sociology at the University of Chicago. By 1908 he had served on multiple commissions for Chicago and Illinois. His interests and advocacy had made him a not-infrequent visitor to Hull House, and he knew Alice Hamilton well through his visits and her recent writing. He had studied the German Sickness Insurance for industrial workers while in Germany and believed the United States should have something similar. A thorough survey of the problems would be a necessary first step. He was also on familiar terms with the reform-minded governor of Illinois, Charles Deneen. In December of 2008, through the influence of Henderson, Governor Deneen created the Illinois Commission on Industrial Diseases. Alice Hamilton was appointed to the commission, also through the recommendation of Henderson.[13]

Almost simultaneous to the creation of the Illinois Commission, the American Association for Labor Legislation (AALL), founded in 1906 to study labor law, was broadening its scope to investigate industrial disease as well. Three early leaders of the AALL played crucial roles in launching the first systematic investigation of a particular industrial poison, the aforementioned phosphorus report. In 1907, John Commons,[14] professor of economics at the University of Wisconsin and contributor to the Pittsburgh Survey, became the secretary of the organization. Commons saw himself more as a theorist than professional organizer and was therefore glad to assign "all the work" to his graduate assistant, John Andrews.[15] Andrews would shortly assume the role of lead researcher of the AALL-sponsored phosphorus investigation. In 1909 he also officially assumed the role of secretary of the AALL. Irene Osgood (Andrews),[16] a social worker by training, served as the assistant secretary of the AALL beginning in 1908 and worked closely with both men.

In early 1908, the new secretary of the AALL, John Commons, sought the advice of several physicians, including Alice Hamilton. His inquiry was about how to begin researching industrial phosphorus poisoning.

Although Hamilton had not conducted any original research, her writing had already established her as worthy of consultation, likely due to the dearth of any experts in the United States. Nevertheless, she declined to give advice, citing her inadequacy to the task. Irene Osgood also wrote to her on December 1, 1908, with a request that Alice Hamilton conduct the phosphorus investigation or, if she declined, that she recommend someone else. Hamilton responded in January 1909 with a letter rejecting the idea that she had any expertise in industrial poisons. In it she declined the offer, insisted that she could offer absolutely no advice on how to investigate industrial poisons, and claimed she did not know anyone who could, except perhaps someone from the Pittsburgh Study or Charles Henderson.

While investigating the timeline of this crucial period, the authors of this book noticed two discrepancies in the various records. First, in her letter to Ms. Osgood, Alice Hamilton claimed to know practically nothing about investigating industrial poisons. There is certainly an element of truth to this. No one in the United States had done such a study. There were no experts. The only completed study, the results of which were beginning to be made available, was the Pittsburgh Survey, in which industrial poisons were only one component. But Alice Hamilton was extremely well read on the topic of European studies of industrial poisons, as her September 1908 article in *Charities and the Commons* demonstrated. According to Hamilton's autobiography she would even have been quite familiar with the phosphorus industry at that time: "Then in 1908 John Andrews came to Hull House and showed me the report of his investigation of American match factories and his discovery of more than 150 cases of phossy jaw."[17]

This led to the second discrepancy. Why was Irene Osgood asking how one should initiate a study of phosphorus poisoning if John Andrews was already conducting such a study? Osgood would have known his work well. Andrews was performing most of John Commons's official functions at the AALL and Osgood was the assistant secretary there. Additionally, Irene Osgood and John Andrews would be married the next year. Although the exact beginning of their romantic relationship is unknown, it is quite possible that they were already involved when the Osgood-Hamilton correspondence took place.

The first discrepancy might be explained by Alice Hamilton's insecurity and modesty. Her insecurity, however, which had been acute in her early years at Hull House, had by this time largely abated. Her relative modesty was more or less a constant throughout her life, although when she had command of the facts she typically spoke directly and with confidence. It is possible that although she was well versed in the current research, the fact that she had never conducted such a study led her to understate her knowledge. The weakness of this explanation is that after her communication with John Commons and immediately before her communication with Irene Osgood, she accepted a position with the Illinois Commission on Industrial Disease. It seems that her humility would have made her just as likely to decline that offer as well.

The answer to the second discrepancy was discovered in John Andrews's phosphorus report. According to Andrews, the United States Bureau of Labor began an investigation of women and children in industry (including in match factories) in December 1908. A few months into their study, they realized that Andrews and the AALL were investigating the same factories, albeit with a different objective, exploring the scope of phosphorus poisoning. They decided to join efforts, and Andrews's investigation thus also became a federal investigation. The comparative timing of the bureau's and the AALL's start dates to their investigations was revealed by R. Alton Lee. The Bureau of Labor began their investigation "just prior to" Andrews's. Thus Andrews did not begin collecting data until early 1909. Hamilton's claim in her biography that he visited in 1908 was off by one year. His visit to Hull House must have occurred sometime in 1909: in 1908 he had no data, and by 1910 the visit would not have been noteworthy, because the report had already been released.[18]

The Illinois Commission, with negligible funding in 1909, limited itself to stating a few basic facts, outlining the scope a comprehensive study of Illinois should take, and describing what would be needed to conduct such a study. They estimated that two years of funding and experts in bacteriology, chemistry, and pathology would be needed. The Illinois General Assembly allocated funding for nine months. In March 1910 Alice Hamilton resigned from the commission in order to accept appointment as medical investigator for the Illinois Survey. Her responsibilities were twofold: supervise the entire project and conduct the investigation of lead, by far the most widely used toxin.

The Illinois Commission had already decided beforehand that the study should be limited to trades in which a known poison was used and the connection between the occupation and disease was already presumed to exist. The poisons included lead, arsenic, brass, carbon monoxide, various cyanides, and turpentine. Although the Illinois Survey was obviously smaller geographically than Andrews's national phosphorus study, the investigation itself was distinctly more complex. Aside from investigating six distinct poisons compared to one, identifying exactly where and how those poisons were being used was far from clear. In contrast, phosphorus was only used in one industry, matchmaking. Andrews's team easily discovered the number of match factories, sixteen, and visited all but one. So although Alice Hamilton borrowed methodological ideas from John Andrews's study, the scope of her task demanded that she develop her own original methods as well.

While the Illinois Survey began with a plan to examine six manufacturing toxins, early visits led them to expand that list (see table 6.1). This is not surprising considering that this was a pioneering study. This was the first time a state had attempted a comprehensive examination of industrial poisons. Many of the industries, such as brass manufacturing, were known to result in illness, physical weakness, and premature aging among the workers. Others, such as photography and glass etching, were discovered along the way. Some manufacturing processes were examined for one toxin, and in the process another was discovered. When brass production was being examined, zinc was also found to be causing illness among workers and therefore was examined both along with brass and independently. And while studying carbon monoxide poisoning among steel workers, they also noticed high rates of deafness[19] (see table 6.1).

The reaction of factory owners to Hamilton's visits was mixed. Many probably believed it advisable to grant the investigators entry for fear of future mandates or to at least appear cooperative. Some were surprisingly transparent—even the leaders of some of the most unsafe factories. Others were intentionally deceitful.

One example of deceit was associated with the discovery of a previously unknown use for lead. Alice Hamilton did not limit her visits to places of work; hospitals sometimes provided the most useful data on illness rates. While she was interviewing a man suffering from colic (severe abdominal pain) and wrist drop (partial paralysis) at Alexian Brothers' Hospital in Chicago, the man claimed his job had been to apply enamel to bathtubs. Hamilton, who knew the German and English literature on

Table 6.1. Comparison of the Illinois Survey and the Bureau of Labor Phosphorus Study

Study	Illinois Survey	Bureau of Labor Phosphorus Study
Official Title	Report of Commission on Occupational Diseases	Bulletin of the Bureau of Labor No. 86
Years Researched	March 1910–December 1910	December 1908– sometime in 1909
Date of Publication	January 1911	January 1910
Publisher	State of Illinois	United States Bureau of Labor
Lead Researcher	Alice Hamilton, MD	John Andrews, PhD
Staff	20 Doctors, Medical Students, Social Workers	Bureau of Labor Members of the AALL
Poisons *Originally only the first six poisons were to be investigated.*	Lead Arsenic Brass Carbon Monoxide Cyanide Turpentine --------------------------------	Phosphorus
As the investigation discovered new dangers, they decided to list other poisons, processes, and symptoms.	Zinc Metol (a salt) Platinum Undetermined (in processing silver for mirrors) Hydrofluoric acid Compressed air Plus deafness, nystagmus	
Industries	Smelting, printing, plumbing, dyes, painting, mechanical art, lithography, batteries, glazes and enamels (lead) taxidermy (arsenic) Brass processing (brass, zinc) Steel production (carbon monoxide, deafness) Photo engraving (cyanide) Turpentine manufacturing (turpentine) Painting/varnishing (turpentine) Photography (metol, platinum) Mirror manufacturing	Match manufacturing

(continued)

Table 6.1. (continued)

	Glass etching (hydrofluoric acid) Construction, various (compressed air) Mining (nystagmus)	
Number of Manufacturing Sites	Unknown	16
Number of Manufacturing Sites Visited by Poison (Or, in a Small Number of Cases, by Resulting Symptom/ Disability)	Lead: 304 Arsenic: 3 Brass: 89 Carbon monoxide: 5# Cyanide: 40* Turpentine: (62 individuals) Zinc: 3 Metol: 40* Platinum: 40* "Silvering" Mirrors: 1 Glass Etching: 1 Construction: (161 individuals) Deafness: 5# Nystagmus: (Approx. 500 individuals)	15

\# Carbon Monoxide poisoning and deafness were both examined in steel works.

* Forty photo studios were visited. All were tested for cyanide, metol, and platinum.

the subject thoroughly, had never heard of this. She visited the factory and inquired with the manager. He assured her that no lead was used on the tubs in the manufacturing process. He showed no defensiveness and willingly allowed her to watch the men painting the tubs. She recorded the name of the paint and verified with the manufacturer that it contained no lead. Puzzled how this man could be obviously symptomatic while no evidence of lead could be found, she tracked him down at his home and inquired further. He explained that she had only seen the finishing process and that the lead application was done at another location. When she located the additional site, she found a highly dangerous process. Unfinished tubs were heated until the metal glowed red. A fine enamel dust was then sprinkled on the tub so it would melt evenly on the surface. The air was sufficiently full of dust that both breathing and swallowing it were inevitable. A workman on site offered to share some of the dust, which he said he often took home to use as an abrasive to clean cooking utensils. Her tests indicated that it contained 20 percent soluble lead.

Alice Hamilton rooted her research in what she had learned from European studies, particularly British and German ones. They provided guidance about the dangers and safety measures to look for in factories and smelting plants, such as lead dust and ventilators. Some of her work confirmed the European literature. Some of it revealed uniquely American challenges to worker safety. Aside from the aforementioned bathtubs and other sanitary ware, lead was found in freight car seals, coffin trim, wrappers for cigars, and decals for pottery decorations. Lead was also used in the polishing of brass and cut glass.[20]

In the midst of her Illinois investigation, Hamilton was selected by the commission to represent it at the International Congress on Occupational Accidents and Diseases in Brussels. Her career thereafter was devoted to researching, reporting, teaching about, and advocating for the reduction of industrial diseases. Years later, Hamilton explained that this transition marked her achievement of a vocation that satisfied all her demands. "I have never doubted the wisdom of my decision to give it [the lab] up and devote myself to work which has been scientific only in part, but human and practical in greater measure."

The Brussels trip was both embarrassing and enlightening. The US delegation comprised just two people, as opposed to the much larger delegations of other nations. More importantly, she found herself unable to answer the vast majority of questions the various European delegates posed to her about lead poisoning in the United States. However, Hamilton benefited greatly from meeting and interacting with the researchers from Germany, England, France, Austria, Netherlands, Belgium, and Italy whose work she had read.

Despite the embarrassing American showing, the conference proved to be a boon for Hamilton's career. One observer at the conference was Charles O'Neill, commissioner of labor in the Department of Commerce. He realized that although Hamilton's knowledge was inferior to that of the Europeans, it was superior to everyone else's in the United States. Shortly after Hamilton returned home to complete the Illinois survey, he invited her to head a similar lead survey for the entire nation.[21] This set in motion a series of national investigations that encompassed multiple industries. This would quickly establish her as the undisputed leading researcher in industrial toxicology.

As her work for the federal government was originally conceived, there were to be several differences between it and the Illinois study.

Alice Hamilton was given much greater autonomy at the federal level. No one would supervise her. Once she finished the lead study, she would move to other toxins or industries, one by one. She would determine how long each study required. However, there would be no salary. Only her expenses would be covered until each survey was completed. Once she finished each report, she would then negotiate the price with the Bureau of Labor.

Alice Hamilton's hard-won experience in Illinois, which initially involved significant guessing and false trails, made the national survey much more efficient. Her productivity in the next few years was striking. She produced one federal lead report each year from 1911 (the same year the Illinois report was released) through 1915 and a final one, in 1919, on women in the lead industries. She began with the manufacture of white lead, the type most responsible for illness and death. In subsequent years she produced reports on lead in various ceramic industries, the painting trades, the smelting and refining processes, and the manufacture of storage batteries.[22] In the same time span, she published eight additional professional articles on the lead industry, along with continued writing about previously researched topics, such as tuberculosis and typhus.

However, counting publications is only the most obvious measure of Alice Hamilton's importance and dedication to reducing deaths and disabling illnesses from unnecessary workplace poisonings. As she explained in her autobiography, "Every article I wrote in those days, every speech I made, is full of pleading for the recognition of lead poisoning as a real and serious medical problem." Her commitment to this cause could be seen in how she adapted her message to each audience she targeted: lawmakers, physicians, industrialists, and the public. For example, Hamilton, in "The Economic Importance of Lead Poisoning," backed her assertions with detailed descriptions of industrial processes, resulting illness rates, and associated financial costs to owners. She argued convincingly why cleaner facilities were not only an ethical necessity, but a smart financial investment.

Alice Hamilton grouped the symptoms associated with lead poisoning into four categories. The first group, general symptoms—by which she meant early, slow-developing symptoms—were largely alimentary. These included loss of appetite, foul breath, indigestion, and constipation. Headaches were also common. Second, workers also often experienced acute attacks of colic (sharp, intense abdominal pains). If a worker quit after experiencing either general or acute symptoms, he or she might make a full recovery. If not, the worker would likely experience chronic effects. These included weight loss; paleness; worsening indigestion and/ or constipation; and diseases of the heart, kidneys, blood vessels, and liver. Many workers died from the internal organ damage. If the worker survived and continued in a leaden environment, he or she would likely experience the fourth category, severe neurological damage. Partial paralysis was the least serious of these symptoms, most often in the wrist ("wrist drop"). Blindness, epilepsy, and insanity were unfortunately common. The epileptic convulsions sometimes resulted in death.

Alice Hamilton described five different lead compounds that are chemically distinct from one another and, more importantly, that differ enormously in terms of danger to workers. Solubility—meaning, in this context, the ease with which lead dissolves in water—is the most important property in determining danger. Red lead (lead [II, IV] oxide, Pb_3O_4), used in batteries; and chrome yellow (lead [II] chromate, $PbCrO_4$) or chrome green lead (chrome yellow mixed with iron blue), which were used in paint, were less soluble and therefore less dangerous. Metallic lead is, by itself, not particularly dangerous. However, when exposed to air at high temperatures, which it inevitably was in many manufacturing processes, metallic lead forms lead oxide (PbO) on its surface. Lead oxide is both soluble and, when formed in many early manufacturing processes, generated a "fluffy" material, so it was easily blown off the surface and floated in the air that the workers breathed. Sugar of lead (lead [II] acetate, $Pb[C_2H_3O_2]_2$), used in making dry paints and dyes, is the most soluble, and therefore the most damaging if ingested. It is, however, very bad tasting, and so workers frequently spit much of it out. The type of lead that disabled or killed the most workers was white lead (lead [II] carbonate, $PbCO_3$). Similar to lead oxide in terms of toxicity and only slightly less toxic than sugar of lead, white lead did not taste bad and was used in many more manufacturing processes than any other type of lead.

White lead was produced through a process variously called the Carter, Stack, or Old Dutch method. This required powdered lead ore to be dissolved in acetic acid, steam, and carbon dioxide for twelve days. One of the danger points in the process was the transporting and dumping of the lead. Powdered lead ore was shoveled into wheelbarrows, which were pushed by the workers, and then was dumped into large cylinders where the workers were exposed to the acid. In both the shoveling and the dumping of the lead, large plumes of dust were created and inhaled. Several times through the twelve-day cycle, any lead that had caked to the sides of the cylinder had to be scraped off by a man who climbed inside to do so. The ore was then melted. Some factories had men transport it to superheated pots, which then created clouds of lead vapor. The newer, safer, but unfortunately uncommon Dwight-Lloyd Method was to place the ore on a grated conveyer belt where it was roasted and melted. As it melted it was sucked through the grate to a large container below. This newer method produced only minimal vapor.

The third and final highly dangerous step was cleaning the liners of the flues. Cotton or wool liners were installed in the flues to trap fine particles of lead and other metals that were economically valuable. As smoke or dust rose in the air, much of it was trapped in these liners. The flues led to a room with bags (the "bag house") of the same material that trapped the remainder of the dust. Men occasionally were required to enter the bag house to shake loose the particles and collect them. This was probably the most dangerous job in the factory because the dust was so thick and the particles so small.[23]

Of all her audiences, the one Hamilton cared most about was the industrialists. If she could convince them to make the necessary changes in their plants, convincing physicians, politicians, and the general public would be both easier and less necessary. Hamilton noted several themes that she observed among industrialists in her investigations. These themes would be observed in later investigations of other toxins as well. There was a frequent tendency to blame the workers for the symptoms, particularly if the symptoms involved fatigue. If the workers were minorities who were already commonly stereotyped, the tendency was even more pronounced. Factory owners were also often ignorant of the harm they were causing workers. This became evident when a surprising number made improvements upon being educated, even without threat of consequences. Among those who resisted change it was often because pride in their work made it very difficult for them to admit to themselves that things were terribly harmful. When laws were passed allowing workers to receive compensation for illness, most of the rest made improvements to avoid paying.

Just as Hamilton noticed several themes in others' behaviors, it is possible to identify several strategies she began to use to increase her chance of effecting change. First, she was willing to work with anyone, including the owner of the "worst white lead factory I have ever seen." Second, in part because of her very limited authority and the strong motivation of some factory owners to disguise the problems, she became adept at cultivating sources to aid her cause, including "apothecaries, visiting nurses, undertakers, charity workers, [and] priests." Third, she realized the importance of understanding the politics of each situation. Fourth, she developed her pattern of documenting the issues and educating her readers that she would use effectively throughout the remainder of her career. This pattern included telling anecdotes of people she met, explaining how the poisoning occurred in a manner that was intelligible to her audience, describing the symptoms and their relationship to one another, detailing the precise industries and processes involved in the poisoning, and reflecting on how various industrialists and politicians responded. Finally, Alice Hamilton would always insist on irrefutable evidence before making her case—the characteristic for which she was best known.

Success was certainly not universal, and how to challenge people determined to oppose her was not an easy choice. When a fight was unlikely to produce desired change, she demurred. On the other hand, peaceful relations with an adversary were not in themselves of value, a lesson learned from Julia Lathrop. On occasion Hamilton would publicly express her frustrations. In an investigation of a phenomenon colloquially named "dead fingers" for the impact using an air hammer had on workmen, Hamilton stated ironically, "It is, of course, shortsighted of workmen to see only the unemployment and misery that threaten themselves and their mates, instead of rejoicing in the wonderful technological advances of the day, but that is the way working people are made."[24]

Given Hamilton's agreement with the Bureau (later Department) of Labor, she continued to investigate various industrial processes at her discretion. One investigation, carbon monoxide, was organized around the toxin. Most were investigations of specific products or industries that utilized multiple toxins. These included rubber; printing; dyes; and her final federal investigation at age seventy-one, rayon.[25]

It is likely that she would have investigated several more industries of her choosing if not for the interruption of World War I. The war shifted both the priorities of the United States government and Alice Hamilton. The government required her to investigate the rapidly expanding munitions industry and later the safety of copper mines in the western United States. Hamilton, by this time an ardent pacifist, struggled with how to balance her antiwar activities with being a contract investigator for the government who first sold weapons and then actively participated in the war. From April to July 1915, Jane Addams led a delegation of over fifty American women to The Hague for the International Congress of Women to discuss how to end the war. Even among many suffragists and peace activists, this was considered a radical action. Alice Hamilton accompanied her as an unofficial delegate, Addams's personal physician, and eventually a chronicler of the congress. This is discussed in greater detail in chapter 7.

Following the government instructions to shift her research to munitions, Hamilton took a brief respite on Mackinac Island and then returned with her normal energy, investigating munitions factories. Unlike her other studies of specific toxins and peacetime industries, the various munitions industries were either entirely new or had expanded so rapidly as to scarcely be recognizable. The presumed toxicity of the manufacturing process was the determining factor in which factories she studied. She did not investigate black powder (gun powder) because although the manufacturing process was clearly dangerous, that danger was entirely from explosions, not poisoning. All other explosives fell under her purview (see table 6.2, which is taken largely from Alice Hamilton's report cited in footnote 22).

As shown in table 6.2, nitrogen oxides were responsible for over half the illnesses and deaths. Nitrogen oxides are a class of chemicals that are composed entirely of nitrogen and oxygen but vary in their exact combinations. Examples include nitric oxide (NO), nitrous oxide (N_2O), nitrogen dioxide (NO_2), and the nitrate ion (NO_3^-). The nitrate ion is commonly associated with other elements, including hydrogen, in which case nitric acid (HNO_3) is formed. Nitrogen oxides were used in a variety of wartime industrial processes. The most widespread and dangerous of these was the use of nitric acid in the manufacture of gun cotton, also known as nitrocellulose.

Gun cotton was a highly useful material because it formed the base of several different explosives that produced less smoke and were more stable than traditional gunpowder. Ballistite, used primarily by the American military, and cordite, sold largely to the British, were but two examples. Although the end products were relatively stable, the manufacturing process was not. That process involved soaking cotton in aqueous solutions of nitric acid at specific concentrations. The exact concentration of the solution depended on the intended use of the gun cotton. The end product needed to be dry and fluffy, and wringing it to remove the nitric acid was a

Table 6.2. Sources of Poisoning among Workers in Munitions Plants

	Recorded Illnesses			Deaths		
Poison	Men	Women	Total	Men	Women	Total
Nitrogen Oxide	1,389	0	1,389	28	0	28
Trinitrotoluene (TNT)	660	43	703	11	2	13
Benzene	14	0	14	7	0	7
Nitrobenzene and Toluene	12	0	12	1	0	1
Ether	52	0	52	1	0	1
Phenol	2	0	2	1	0	1
Mixed Acids	2	0	2	1	0	1
Sulphuric Acid	4	0	4	0	0	0
Picric Acid	7	0	7	0	0	0
Aniline	203	0	203	0	0	0
Fulminate	79	32	111	0	0	0
Ammonia	1	0	1	0	0	0
Mercury	1	0	1	0	0	0
Nitronaphthalenes	2	0	2	0	0	0
Chlorine	3	0	3	0	0	0
Total	2,433	75	2,508	51	2	53

standard procedure to achieve that. It was important to carefully control the conditions during the manufacturing of gun cotton. If the necessary conditions were incorrect, the chemical composition could change rapidly and possibly result in a violent chemical reaction that sprayed a fine mist of acid, resulting in burns to both the skin and lungs.

The symptoms of such exposure varied. Skin exposure resulted in severe burns. Mild cases of inhalation resembled asthma attacks from which workers could recover. Heavier exposure produced symptoms that closely resembled the symptoms of poison gas attacks in the war, including death.[26]

Hamilton's independence in conducting her investigations was accompanied by a lack of formal authority to visit sites and very little official guidance in where to look. Somewhat surprisingly, this did not change during her munitions investigations. "If there was anyone in Washington who knew where explosives were being produced . . . he kept the secret." Dr. Meeker, in the Department of Labor, to whom she reported, suggested she simply use the methods that had served her well before, such as visiting hospitals, bars, union halls, and people's homes as well as following up on rumors. Once she knew the general location of a factory, she was aided in the process by accidents at the plants themselves. Giant plumes of orange smoke would sometimes guide her to the factory, and the location was confirmed by the sight of dozens of "canaries" (men covered with yellow acid).

Hamilton recalled her first visit to a munitions factory with some amusement. This was a gun cotton plant on the James River, south of Richmond, Virginia. Unlike many other factories, its exact location was not a secret. The obstacle in this case was of a very different nature. When Alice Hamilton arrived in town, she was accompanied by her friend, dating from her time in Miss Porter's School, Mabel Kittredge, who had become a major voice for school lunch programs. They had a very difficult time finding a place to stay and noticed that police were paying them an unusual amount of attention. They were fortunate that the Greek restaurant they chose for dinner happened to be run by an old friend she knew from Chicago. He explained to them that the town was almost entirely devoted to serving the interests of the factory workers and as a result the only women in town had been prostitutes. Not long before, an evangelist had made it his mission to clear the town of prostitutes and had convinced the mayor to join him in that effort. Once the evangelist had them removed, the town was entirely masculine. Hamilton, at that point forty-eight years old, was bemused and somewhat flattered to be mistaken for a woman of the night. By this point in her career, she was accustomed to being direct, so she made it a point to meet with the plant manager (who agreed to explain it to the police) as well as the evangelist to debunk the rumor that she and Mabel were prostitutes. In both cases she described fruitful and interesting conversations, and she had no further trouble with that investigation.

On April 6, 1917, the United States entered the war. A follow-up munitions study would be ordered shortly afterward, and in it, Hamilton would report safety improvements in both gun cotton and picric acid plants. However, just a month earlier, the unusual health condition known as "dead fingers"—spastic anemia of the hand—came to her attention. Her investigation into that condition would both divert her attention from the war industries and eventually point back to them.

On March 19, 1917, a fellow physician from Chicago, Joseph Miller, wrote to her describing his patient, Mr. McBrairdy. Mr. McBrairdy was a stonecutter in the southern Indiana limestone quarries. His symptoms included intermittent numbness and blanching in his left hand. He also told of coworkers who experienced similar symptoms, some of whom also complained of pain and weakness in their entire left arms. Dr. Miller speculated in his letter that this was an unusual type of Raynaud's syndrome, a condition marked by pain; numbness; and loss of blood to hands, feet, nose, ears, or lips—caused by cold temperatures or emotional stress.

The likely connection between Mr. McBrairdy's symptoms and work as a stonecutter was not lost on Dr. Miller. Stonecutting was obviously an old profession, but these symptoms were newly discovered. However, a new device had been introduced approximately twenty years earlier, namely the air hammer. This prompted his letter to Hamilton.

The Bureau of Labor opened an investigation shortly thereafter and sent two agents. It is unclear whether their involvement came from a referral from Dr. Hamilton, but considering Hamilton's close relationship with the bureau, the timing of Dr. Miller's letter, and that the investigation began in southern Indiana, it is likely that she played some role. However, there were other concerned voices. The Journeymen Stone Cutters' Association, along with John Andrews's AALL, were both demanding an investigation.

The early investigation produced contradictory information. The bureau agents came to Indiana to gather statements from both stonecutters and their employers. Further information was provided by a Dr. Cottingham, who was employed by the union. Dr. Cottingham, along with most of the stonecutters, claimed that the use of the air hammer not only damaged the hands of the workers, but that it led to tuberculosis (presumably from inhaling large amounts of limestone dust), paralysis, neurasthenia (fatigue and irritability), and even insanity. The employers, along with nearly a third of the workers, claimed that it did little or no damage at all.

The air hammer resembled a small jackhammer but was held differently. Its design required the stonecutter to hold the body of the tool with the right hand (for right-handed workers) and the chisel with the left. The chisel delivered approximately fifty strokes per second.

Alice Hamilton was sent in later that spring to sort through the contradictions and determine what impact, if any, the air hammer had on the workers. After a brief visit she decided to return during the winter, when the symptoms were reportedly more severe. That winter she also visited granite cutting centers in Massachusetts and Vermont, marble shops in Long Island and Baltimore, and sandstone mills in northern Ohio.

Symptoms varied among the workers, almost entirely based on how much they used the air hammer (see table 6.3). Limestone workers used the air hammer for all their work (see table 6.3) and subsequently suffered the most. One Indiana worker's left hand was greenish-white in parts of all four fingers, including the entire little finger. He rubbed his hand vigorously or swung his arm to ease what was otherwise unbearable pain. When blood had returned to the hand, there was a stark delineation between the white, corpselike fingers and the swollen redness of the remainder of his hand. The right hand was also symptomatic, albeit less severely. The marble and granite cutters suffered similarly to one another. Marble cutters used hand chisels occasionally, but most of their work was done with smaller air hammers. The granite shops provided the clearest evidence that the air hammer was the sole culprit. Those who cut the blocks (cutters) did not use the air hammer and were asymptomatic. The carvers (e.g., those who did lettering for tombstones) used them regularly and suffered symptoms similar to the marble workers. Those who worked with sandstone knew nothing of dead fingers since many did not use the air hammer, and those that did used it infrequently.

In this investigation Hamilton again had the opportunity to demonstrate her objectivity and honesty. As discussed in detail in chapters 5 and 7, she gradually moved toward more radical political positions, driven largely by her exposure to social and economic inequalities. However, she was not an ideologue, and her belief in truth-telling overruled any biases she may have had. Her report supported the union and most of the workers in their assertion that dead fingers was a real phenomenon that caused severe pain and eventual disability. However, she found no support for the stories of tuberculosis and insanity in the Indiana limestone quarry and clearly reported that. She did, however, find extremely high rates of tuberculosis among granite workers. Multiple men at the Vermont site told her directly that they expected to die from lung disease, which they did at a rate sixty times higher than the general population.[27]

Table 6.3. Working Conditions and Use of the Air Hammer across Types of Stone

Stone	Use of Hammer	Winter Conditions	Symptomatic Men
Limestone (Indiana)	Cutters used for all cutting with various sized hammers.	0° Celsius	34 of 38
Sandstone (Ohio)	Half used it occasionally. Half did not use it.	0° Celsius	0 of 15
Granite (New England)	Cutters did not use it. Carvers (e.g., letters on tombstones) use it 4–6 hours/day.	Below 0° Celsius	43 of 50
Marble (Mid-Atlantic)	Cutters used for most cutting, usually with smaller hammers.	Below 0° Celsius. Better sheltered than granite.	44 of 78

During the investigation among stonecutters, events in the Southwest were unfolding that led to her final war-related investigation. Cochise County, Arizona, and the surrounding area produced large amounts of copper. During peak production in 1917, over thirty million pounds of copper were mined each month. This accounted for approximately 28 percent of copper production in the United States during World War I. Not surprisingly, the federal government viewed any threat to that production as a matter of national security.

In June 1917 tensions were high between mine owners and miners. Safety, hours, and salary issues were central, but the situation was more complex than that. The majority of the owners were autocratic and antilabor. As such, they could coerce more flexible or compassionate owners not to negotiate or improve working conditions. The workforce was transient, making organization difficult. There were also racial tensions between English and Spanish speakers. A growing trade union movement among the miners had made some progress in unifying the workforce, but the divisions remained. Mining was so much the dominant industry that nearly everyone had taken sides in the growing conflict. Doctors, lawyers and preachers were all known to be pro-owner or prolabor.

On June 26, 1917, the union called a strike in the Warren District of Cochise County, near Bisbee, effective the next day. The mining company, Phelps Dodge, had no designated negotiators to discuss terms with the union, making conflict and force the natural response to perceived threats. The strike was nonviolent. Nevertheless, Sheriff Harry Wheeler requested federal troops to end it. An army officer visited on June 30 and July 2 and found nothing illegal or dangerous. No troops were sent.

Two weeks after the strike was called, the mine owners, along with the sheriff, met to plan a course of action. On July 12, the sheriff, leading a force of approximately 2,000 armed men, rounded up 1,186 miners, forced them onto a train, and transported them 170 miles to Columbus, New Mexico. The authorities in Columbus refused to let them leave the men there with no provisions, and so on the return trip, they left them in the desert town of Hermanas, New Mexico. The men were left there for two days until the army discovered what had happened. Soldiers were sent to transport the men back to Columbus, where they were cared for until they could be returned home.

A presidential commission sent to investigate filed two reports, each placing the blame on the owners of the Phelps Mine and the sheriff's department. Federal mediators sought to resolve outstanding disagreements between the workers and Phelps. Several perpetrators were arrested, but none were convicted because federal law at that time did not apply.

A year later, resentments and mistrust remained pronounced. Labor leaders assumed that any dangers to which the men were exposed would be hidden by management. When they learned of Hamilton's ongoing investigation of dead fingers among stonecutters, they speculated that the use of jackhammers in the mines might have a similar effect and requested an investigation. The federal government, sympathetic to the miners following their kidnapping by the sheriff and desirous to keep them contented and working, sent Alice Hamilton to investigate.[28]

Several issues converged in the Arizona copper mine investigation. The miners were concerned about dead fingers, another topic for which Alice Hamilton was now the foremost expert. The federal government wanted to keep the mine open for the war effort and therefore took the miners' concerns very seriously. Finally, labor and management each mistrusted the other. Hamilton instinctively sided with labor but also wished to cultivate positive relationships with management as well.

Considering all that was at stake and the tensions resting just below the surface, the investigation itself was arguably the least revealing of her career. Alice Hamilton visited at least seven mines during her trip. She interviewed men at each mine and, by her request, saw them using the jackhammer in a wide variety of situations. Some of the men complained that using the jackhammer made them nauseous, which was not surprising because it was often necessary to hold it against the abdomen for maximum effect. But she found that it caused no lasting harm. The jackhammer was much heavier and vibrated much less, and holding the chisel in one's hand, the primary cause of spastic anemia, was out of the question.[29]

Although no permanent damage appeared to result from the jackhammer directly, she did note that it created an excessive amount of dust. In some mines, water was used to wet the rocks and reduce dust, but not in nearly as many as was ideal. Management explained that they made the water available to the men who often chose not to use it. Many of the men countered that they would gladly use it, but the extra time it took to wet the rocks was not counted as work time, so they were being penalized for taking safety precautions. Hamilton also noted that among Mexican workers there was a strong desire to avoid getting wet. She suspected that the dust might contribute to tuberculosis, but since her assignment was limited to spastic anemia, her suspicions were only recorded in her unofficial notes and later in her biography.

She also noticed that at least one man lost up to forty pounds after working near the acid baths used to separate the copper from the tailings (waste products). Since copper ore often includes trace amounts of arsenic, she suspected comparatively mild arsenic poisoning in the fumes as the culprit. This too was outside her assigned investigation and was therefore left as speculation.

Alice Hamilton managed to establish trusting relationships with both management and labor. She was even able to broach the topic of the Bisbee deportations with one of the engineers. He had arrived after the deportation and saw it as a poor decision, but his sympathies were with management, as was the attitude of most of the skilled workers. He justified the actions of the sheriff's men as "getting carried away" and pointed out that no harm came to the men, even though it certainly would have if the army had not discovered their location.[30]

Hamilton conducted no new investigations in 1919, the year she received her invitation from Harvard University. She was, however, very busy writing and traveling. She and Clara Landsberg arrived in Chicago from the Arizona mines in early January. The next few months were spent (1) authoring a report on the advisability of the state of Illinois providing public health insurance, and (2) publishing reports on lead and various munitions. In April she sailed for Europe with Jane Addams, Florence Kelley, Jeannette Rankin, and others for the Second Women's Peace Conference and to coordinate relief efforts in Germany (see chapter 7). Alice Hamilton must have spent much of her spare time writing based on her multiple tomes about the journey, her ongoing writing commitments about recent investigations, and simultaneously preparing to become the first female professor at Harvard.[31]

SEVEN

THE SCIENTIST AS SOCIAL SCIENTIST

ALICE HAMILTON LEARNED ADVANCED SCIENCE in medical
school but advanced social science—and humanities—at Hull House.
This chapter explores her evolving views and actions to ameliorate social
problems related to childbirth, child-rearing, sexuality, religion, peace,
economic inequality, and women's rights. Much of this change took place
during the Hull House years (1897–1919), but she never became static in
her views. They continued to evolve during her entire life.

Aside from her highly personal vocational struggles, the most significant change in Alice Hamilton's thinking from 1897 to 1919 involved a
steady shift toward more progressive religion and politics. The primary
catalyst for this shift was her exposure to the various ideas expressed by
Hull House visitors, both humble neighbors and some of the most prominent thinkers and politicians in the world. She carefully considered new
ideas and invested not only her intellect but her entire person into coming
to resolution on important political, religious, social, and ethical issues.

During her early years at Hull House, the flurry of new ideas was dis-
orienting and uncomfortable for Hamilton. Her skepticism, a conservative tendency in this context, meant she would not embrace new ideas
without careful consideration, testing, and time. Her openness, the liberal
counterpart, meant she could not easily dismiss them. The tension between the two, which would serve her so well in the future, was largely a
source of pain in the beginning. Nearly a year after arriving at Hull House,
in a letter to Agnes, she described the tension between skepticism and

openness and her subsequent efforts to maintain her equilibrium in the midst of it: "I am holding on to myself to keep from toppling off the fence which is my only safe and comfortable place."[1]

Eventually Hamilton embraced several principles that would enable her to resolve difficult issues for herself, or at least come to peace with the uncertainty that sometimes accompanies searches for answers to big questions. Those principles included the aforementioned skepticism and openness. Honesty and following the evidence rounded out the list. In other words, she applied a scientific approach both to her work and to her search for truth. For Alice Hamilton that search was ultimately a practical one, because to her, truth was a guide to action.

Alice Hamilton's actions fit a discernable pattern: research with clear practical benefits for people in the greatest need took precedence over specialization or prestige. The investigations into typhoid fever and the sale of cocaine, as well as her eventual investigations of industrial diseases, fit that pattern. A lesser-known study was her research into the impact of excessive childbearing on infant mortality. By surveying her patients at Hull House, Hamilton found a clear pattern of higher rates of infant deaths among larger families. When comparing families with four or fewer children to those with six or more, children in large families were more than twice as likely to die in infancy compared to the children in the small ones. This pattern was even more pronounced in very large families, with the mortality rate increasing as each new child was added to the family. Although tradition and religious doctrine (e.g., opposition to contraception) would make it more difficult to convince mothers to have fewer children, Hamilton followed the evidence and searched for actionable solutions to longstanding problems.[2]

Hamilton's discussion of whether to share new research findings with the general public illustrates the importance of honesty as a guiding principle. In the late 1800s and early 1900s, prostitution was thought to increase the likelihood of contracting tuberculosis. Not surprisingly, this danger was one of the primary arguments used to warn women away from this profession. In her article "Prostitutes and Tuberculosis," Alice Hamilton summarized recent findings that showed little or no relationship between the two. A debate appears to have arisen within the scientific and medical community about whether to publicize these results. Those assuming the negative position argued that without the fear of this disease, women would not be sufficiently deterred from this career choice.

Hamilton argued, "If there is no way of deterring a girl from a life of shame save by terrifying her with the prospect of an early death, then we must give up trying to deter her." The cost of misleading them would be a loss of credibility and thus a loss of the ability to inform and help in the future. Honesty was both right in itself and a better long-term strategy.

Although Hamilton continued to value tradition, new evidence on an issue consistently took priority. In the United States and England, there was a long-held consensus that boiled cow's milk for babies was indigestible, negated the benefit of colostrum, and caused constipation. However, it had been used for many years in continental Europe. After studying the issue carefully, she concluded that although it could, at times, lead to constipation, the other two claims were false. Moreover, in an era where bacteriological contamination was common, the safety it provided more than compensated for the occasional discomfort. Once she reached this conclusion, she publicized this idea both to her clients at Hull House and to the larger public, regardless of widespread skepticism.[3]

While at Hull House Alice Hamilton was quite aware that she was gradually moving away from the pietism of her upbringing toward a religion centered around social justice. She increasingly saw more value in doing justice and showing mercy, and far less in evangelizing. This was quite difficult for her because it created ideological and perhaps emotional distance between Agnes and her. It also led to awkward and, for Alice, embarrassing encounters. One particularly uncomfortable visit involved Agnes and Esther Kelley, the leader of the Lighthouse, a settlement house in Philadelphia. In contrast to Hull House, the Lighthouse insisted on stricter behavioral rules and overt evangelism. When Agnes asked Alice about Ms. Kelley visiting, Alice assumed an apologetic tone about the Catholicism and roughhousing she might encounter at Hull House. She was particularly worried about conflict between Esther Kelley and the sometimes-blunt Ellen Gates Starr, cofounder of Hull House, because Jane Addams and Julia Lathrop would be gone and thus unable to smooth over their differences.

Author Eleanor Stebner described Hamilton as arriving at Hull House a pilgrim and departing a mature reformer, implying that social reform was eventually the primary expression of her faith. She viewed Alice Hamilton's faith as similar to that of her two colleagues, Florence Kelley and Julia Lathrop. For each, most doctrinal statements were viewed with caution, if not suspicion. This caution came more naturally for Kelley, whose

Quaker background held many Christian doctrines as optional. Lathrop and Hamilton would continue to embrace the ethics of their evangelical beginnings, even if not much of the theology. "Love, service, care, and friendship" were the centerpieces of their faith. Action on behalf of people in need was their calling. The motivation of all three women generally included both an ideal that the world was improvable and a belief that they could individually make a difference through lives of active service.

Alice Hamilton referred only sporadically to her faith in her later writing, although the comments that have survived indicate that she continued to wrestle with religious issues. Shortly before her 90th birthday, with Margaret's expected death looming, Alice was more preoccupied than normal with the issue of life after death. In a letter to long-time friend, Katy Bowditch Codman, Alice shared her view of the afterlife.

Somehow, I do not know where or when, I lost my belief in individual immortality. I don't mean I disbelieve it, I simply do not know. When death comes, I shall know. "Now we see through a glass darkly, but then face to face. Now we know in part but then we shall know even as we are known." That is from St. Paul. My whole relation to God and to Christ has become cloudy, and think will be till I die and then really know.[4]

For Alice Hamilton, "love, service, care, and friendship" were not only interpersonal, they were international. As such, she somewhat naturally embraced pacifism. She was a firm opponent of all war from at least the middle of World War I until sometime during World War II. The exact point of her conversion to pacifism is not entirely clear. What can be shown with greater clarity is that her enthusiasm for peace efforts increased dramatically during her passage across the Atlantic to attend a women's peace conference in Europe. Her letter to Agnes on April 5, 1915, shortly before her departure, indicates embarrassment for her decision to attend. Her letter to Mary Rozet Smith, less than three weeks later, conveys confidence that the American delegation had become a coherent and strategic body and that she was enthusiastic about participating in something worthwhile. It is difficult to tell whether her embarrassment in the earlier letter was due to familial disapproval, the general skepticism in the United States about the women's conference, her own doubts about the efficacy of such a conference, or ambiguity about pacifism generally. Whatever her reservation may have been, she returned home without it.

Table 7.1. Major International Travel

1. Germany	Advanced Study	1895–96
2. Belgium	International Congress on Occupational Accidents	1910
3. Netherlands	Peace Conference	1915
4. Switzerland	Peace Conference	Spring 1919
5. Germany	War Relief	Summer 1919
6. Switzerland	League of Nations	Fall 1924 and later
7. Russia	Public Health Service of Soviet Russia	Fall 1924
8. Netherlands	International Congress on Occupational Accidents and Diseases	1925
9. Germany	General Research	1933
10. Spain and Morocco	Vacation	1936
11. Mexico	Vacation	1937
12. Germany	International Congress of Occupational Accidents and Diseases	1938
13. Guatemala	Vacation	1948

The following year, in an article discussing the newly bequeathed rights of women to vote (in some states) and how she would use her rights, Hamilton examined each critical policy issue as she saw them and compared the two main candidates, incumbent president Woodrow Wilson (Democrat), and his challenger Justice Charles Evans Hughes (Republican) on each of them. Ultimately Hamilton's decision came down to two issues: Hughes had endorsed women's suffrage nationally; Wilson had not. But Wilson favored free trade, whereas Hughes was a protectionist. Hamilton saw protectionism as a "breeder of wars," and that appears to have taken priority over any other single consideration. Peace was of even higher value than women's right to vote.[5]

Alice Hamilton made three trips to Europe during her Hull House years (see table 7.1). The first was to participate in the International Congress on Occupational Accidents in Brussels in 1910. Her latter two trips were both associated with peace efforts. In 1915 Jane Addams was invited to preside over an international Congress of Women, organized with the intention of promoting an end to the fighting in Europe. As mentioned earlier, Hamilton was skeptical of what the congress could accomplish.

Nevertheless, she accepted the opportunity to observe, participate, and serve as Jane Addams's traveling companion and personal physician. In spite of her skepticism, she played a not-insignificant role as a researcher and traveling representative of the congress, and even served as an author of a report on its proceedings.[6]

The third trip was to the Second Congress of Women, also with Jane Addams. The second congress, in 1919, coincided with the negotiations to end the war, and as the details of the Versailles Treaty were released, the congress was the first political body to criticize them. The conference was not the only reason for the visit. Future president Herbert Hoover, at that time head of the American Relief Administration, had granted both Addams and Hamilton permission to travel with and assist the Quakers in distributing food to Germans, who were suffering from a severe famine. In the four years since the first congress, Hamilton had evolved from an interested but somewhat peripheral participant to a fully engaged researcher and advocate, determined to do all she could to alleviate suffering and prevent another war.

At the conclusion of the third trip, Alice Hamilton coauthored reports both for the Quakers and in publications that shared an interest in soliciting donations to help feed Germans who were severely malnourished. These reports targeted at least two specific audiences with multiple distinct goals. The first audience was the organizers of food relief efforts, including the aforementioned Herbert Hoover. She provided them with crucial statistics about food supplies and how they translated into calories and other nutrition per person, particularly children.

The second, broader audience was provided with stories and information to encourage donations in order to avert a growing crisis. In her reports, Hamilton sought to use language that appealed to generosity, not guilt. The subtitle of the Quaker report, for example, was "An Eloquent Appeal to the Hearts of Generous Americans." Americans had experienced years of propaganda convincing them that war with Germany was necessary. To now convince Americans that Germans were human beings who were suffering acutely was not an easy sell. One way to accomplish this was to focus on children. For example, German children were subsisting on one-third the calories needed. Such statistics were bolstered by explanations as to why the problem continued. French-occupied German farmland was being used to feed the French. The naval blockade, still in effect, forbade Germans from fishing in the Baltic and North Seas.

For readers less interested in politics and statistics, Hamilton provided anecdotes such as stories of a twelve-year-old girl who weighed forty-six pounds, a boy with rickets who risked breaking bones by simply playing outdoors, and elderly people whose only food for months was turnips.

Alice Hamilton further sought to humanize the Germans in the eyes of Americans by sharing their perspectives and stories. In her short essay "On a German Railway Train," she explained that many of the people who questioned her after her trip asked if Germany had learned a lesson or had realized it was beaten. Her response was that she had not met Germany; she had met Germans. Those Germans varied in their opinions, just as Americans did. She then recounted a lively discussion she had observed on a train. This discussion involved her traveling companion, who was an outspoken Dutch woman, and six Germans, including four former soldiers, one elderly man, and one woman. This format also allowed Hamilton to voice German complaints about their mistreatment without her needing to judge the legitimacy of the claims. Thus, she could introduce both the German people and their suffering in a way that American audiences would hear.[7]

Hamilton explained many of her pacifist ideas and their rationale in a chapter entitled "Because War Breeds War," which she contributed to "Why Wars Must Cease," a 1935 critique of war whose authors also included Jane Addams, Eleanor Roosevelt, and Carrie Chapman Catt. Her ideas in 1935 were shaped substantially by her multiple trips to Europe, most notably by comparing the changes in Germany that took place between the Second Congress of Women in 1919 and Hitler's Germany in 1933. She explains, "Those of us who saw Germany in the early days following her defeat and humiliation can testify that at that time her mood was not one of hatred and vengefulness, but one of desperation and bewilderment." Hamilton explains that there was an excellent opportunity at the end of World War I to disrupt the cycle of one war laying the groundwork for the next. The German people had gone to war because they understood it to be their duty. In general, they bore little resentment for their enemies. In contrast, England and France, as democracies, needed to generate public support. They did this by effective propaganda that left their populations seething. This was particularly true in France, where a significant portion of their nation had been occupied or physically laid to waste. The attitudes of the French could also be traced to their defeat in the Franco-Prussian War nearly two generations earlier. The English and French resentment, which carried the day in the negotiations among the victors, drove the harsh terms of the Versailles Treaty. That treaty, in combination with the inability of the German economy to feed its people, transformed German bewilderment into resentment and a desire for revenge. When a nation has been humiliated, it can no longer be objective and accept realistic blame. This humiliation pushed much of the German populace to embrace Hitler and Goebbels, who borrowed and perfected the Allies' tool of propaganda. Even as early as 1933, Alice Hamilton both saw how the cycle was continuing and predicted it would lead to another war.[8]

Alice Hamilton did not embrace the labor movement as easily as she had pacifism, although she later stated, "At Hull House one got into the labor movement as a matter of course, without realizing how or when." Hamilton was raised in an environment that supported more conventional progressive social movements such as women's suffrage and, more comfortably, temperance. These were more easily combined with upper-middle-class Protestant Christianity. The labor movement was more associated with Catholicism or, more problematically, with anarchism, and embraced rhetoric that was often exaggerated or violent. While she naturally sympathized with their complaints about poor working conditions, her upbringing, particularly the influence of her father, made her leery of organized labor. She described her family's perspective as "right wing liberal," and it would take a preponderance of evidence for her to adopt many of labor's goals, and some of their methods, as her own.

Part of her resistance to the labor movement was a confidence in official authorities. Because most authorities lined up against organized labor, supporting labor felt like disrespect for both business leaders and the police. The Chicago Haymarket Riots were only eleven years past when Hamilton arrived at Hull House. She was therefore uncertain how to mentally process meeting former governor John Altgeld, who had pardoned three men associated with the riot. Trust in the authorities played a critical role in initially preventing Hamilton from embracing the labor movement and socialism.[9]

It is not surprising that evidence of corruption and brutality among police and government officials would eventually facilitate the transition in her thinking. That evidence would mount quickly. During her first year at Hull House, two Italian workmen, whose crime was to refuse to move off of their own garbage boxes, were shot by a police officer. She arrived just moments later, to see a mob forming. The policeman was safely transported from the scene. The workmen were taken to the hospital, where one died. Her growing disillusionment with the entire system was exacerbated by the lack of any penalty for the policeman. It was also during her first year that she observed and played a minor role in attempting to unseat Ward Boss Johnny Powers, an official widely known for encouraging bribes and neglecting the needs of those who could not pay. By the early 1900s, she was more comfortable taking an active role in advocating for laws to protect people in her neighborhood. The threat this time came from the pharmacists discussed in chapter 6, who realized they could make a great deal of money by encouraging addiction to cocaine among young boys and then selling the drug illegally. The legislation she helped pass was hailed as a substantial triumph among reformers but was declared unconstitutional based on a technicality. She thereafter viewed courts with much suspicion as well.

Her shift toward socialism, which continued to evolve throughout her long life, was not marked by such visceral and disturbing events. Rather, it developed gradually through conversations with a wide range of thinkers and activists, most of them visitors to Hull House. Alice Hamilton was methodical in her skeptical consideration of new ideas. Skepticism, to Hamilton, meant setting aside assumptions as best she could and subjecting all ideas, both familiar and new, to careful consideration. That consideration was always to be guided by evidence, which usually had a utilitarian bent. Whatever improved the quality of life, especially for those in greatest need, was probably right.

Hamilton learned that socialists were certainly not a monolithic group. She was particularly disappointed with an unnamed "famous" English socialist, about whom she reflected to herself, "You may love humanity but you certainly do not love your fellow man" when he lectured her on the needs of poor English children while simultaneously pushing poor American children aside as he and Hamilton walked. Ideologically she disagreed particularly with socialists who eschewed democratic methods. In contrast, she spoke very highly of Peter Kropotlew, a Russian Menshevik who advocated both political and economic justice for all and, importantly for Hamilton, evidenced his view of the value of all people through his courtesy to all he met.[10]

As mentioned earlier, women's rights were not at the forefront of Alice Hamilton's agenda; however, there is clear evidence that they were a concern of some importance to her. The March 7/8, 1911, edition of the *Chicago Record-Herald* included an article entitled "Suffragists storm capital (Springfield) with Logic." Eighty-seven people were listed by name, Alice Hamilton among them. This is telling evidence, considering her dislike for attending public protests.[11]

Of the women's rights issues Hamilton embraced, perhaps the most surprising, considering her family background, was birth control. But perhaps this is not so surprising given her medical education and her longtime exposure to large families living in poverty where women were overwhelmed with their responsibilities. Regardless, the gradual but thorough shift in her thinking took Hamilton herself by surprise. She explained, "It must have been some ten or twelve years after I came to live in Hull-House that I was suddenly asked one day in public whether I believed in birth control. Without stopping to think I answered at once that I most certainly did and then realized that this was the first time the question had ever been put to me or I had ever formulated my belief even to myself. The answer had been almost automatic, prompted by my daily experiences in a poor community." Hamilton would later expound on her views in a letter to the League of Women Voters. Using the Garden of Eden story, Hamilton highlighted the first two curses placed on humankind: the curse of manual labor for men and the curse of childbearing for women. She then drew a distinction. Men devoted enormous creativity and energy to reducing the first burden. In contrast, when women complained about the second burden, society, often with the encouragement of the church, forbade easing the curse, giving the pain a sometimes-sacred status. She additionally argued that birth control would lower the number of abortions, a practice of which she disapproved.[12]

This book focuses primarily on Alice Hamilton's personal and professional evolution prior to joining the Harvard University faculty in 1919. However, a glance at Hamilton as she transitioned to Harvard and Hadlyme shows her not only continuing her research on industrial workers, albeit at a slower pace, but maintaining and perhaps accelerating her advocacy for other disadvantaged groups. While at Harvard, and even more so in her retirement, Hamilton corresponded regularly with a variety of government officials and advocacy groups. The themes of greatest importance to her can be inferred from that correspondence, namely peace, freedom of expression, the rights of foreigners, and social/economic justice—themes explored further in chapter 8.

EIGHT

—ᘒᘒ—

EPILOGUE

The Senior as a Public Intellectual

WHEN AT A VERY HEALTHY age sixty-six, Alice Hamilton retired from Harvard—because of "old age," her official personnel record reads—she had more research to pursue, many more years to live, and much more learning to acquire. Her last major research, conducted for the Department of Labor, was an investigation of the viscose rayon industry. She continued to publish through her seventies, producing her autobiography, *Exploring the Dangerous Trades*, in 1943, and reissuing, with Harriett L. Hardy, her text on *Industrial Toxicology* in 1949. Hardy testified to Hamilton's continuing physical stamina. When the two women were visiting research sites that required climbing steps, the nearly eighty-year-old Hamilton would "hop up the stairs as if she were about sixteen or seventeen and I'd come panting behind her."[1]

In 1938 she visited Nazi Germany for the second time. Although this trip was to attend the International Congress of Occupational Accidents and Diseases, the most significant thing that she learned of was not the latest developments in industrial health but the frightening effects of the full-blown Nazi regime on the German people and culture. With another world war threatening, this trip kindled memories of Hamilton's two earlier visits to Europe as a peace advocate during and just after World War I. By the late 1930s she was winding down her career in industrial hygiene and revisiting and expanding her earlier interest in public affairs. More than ever before, the scientist was focusing on the social sciences. Her interest in closely following public affairs and commenting on the same as a public intellectual enlarged during World War II and the Cold War and became her primary "vocation" during her advanced years.[2]

As this interest in public affairs eclipsed that of her interest in public health, the aged Hamilton read widely and acted on her understanding.[3] She regularly wrote appeals, signed petitions, and wrote letters to public officials and newspaper editors. Her major concerns were (1) for conflicts to be resolved peacefully and by dialogue, and (2) for governments to encourage the maximum possible amount of freedom of conscience for their citizens.

In international affairs she favored the United Nations in the same way that she had supported the League of Nations, and she opposed the Vietnam War. Alice Hamilton departed from her traditional pacifism to support the military effort against Hitler in World War II. Thus she became more nearly a quasi-pacifist than an unqualified pacifist.

During the Cold War, Alice Hamilton did not defend Russian Communism but did criticize American efforts to oppose it by restricting freedoms at home or by supporting overseas anticommunist regimes that were nondemocratic themselves. The defense of domestic civil liberty became Hamilton's greatest public cause during the last decades of her life. Concurrently, determining the limits of the Bill of Rights freedoms of speech and assembly became the primary federal government domestic issue of the Cold War era.[4]

Hamilton's protests, as recorded by Barbara Sicherman in her letters of Alice Hamilton, were the most regular in the "coldest" decade of the Cold War, namely 1945–55. It is this period that witnessed (1) the strongest enforcement of the Smith Act (1940), which made it a crime to advocate the overthrow of the government by force; (2) the adoption of the Taft-Hartley Act (1947), with its noncommunist oath for union officers; (3) the McCarran Internal Security Act (1950), which required the registration of communist organizations; and (4) the Communist Control Act (1954), which outlawed the Communist Party as a legal entity. In similar spirit, the Supreme Court approved such restrictive decisions as *American Communication Association v. Douds* (1950) and *Dennis v. the United States* (1951), and the Truman and Eisenhower administrations instituted loyalty programs for Executive Department employees.

Considering that most of this correspondence occurred at the onset of the Cold War and the Red Scare, it is not surprising that the Federal Bureau of Investigation opened a file on her. That file included documents from 1942 through 1965. Her file contains letters from Hamilton to elected officials defending the rights of foreigners who had been treated rudely or worse, as well as letters encouraging officials to vote against legislation that would curtail civil liberties. The file also contains correspondence amongst elected officials and the FBI about how to respond to her and debating how much of a threat she was. There are also lists of organizations to which she belonged or worked collaboratively, and accusations that she was a communist, although "communist sympathizer" is more accurate. The latest item from Hamilton is a letter dated July 15, 1965, when she was ninety-six. In perhaps her final protest, "a very old woman" who had "lived through the McCarthy Era" made clear the immorality of the Justice Department's revived persecution of people based on political beliefs.

After 1955 the mood in the country changed with the end of the Korean War; the death of Russian premier, Joseph Stalin; and the demise and reaction to the red-scare tactics of Senator Joseph McCarthy. This change found expression in the more tolerant Supreme Court decisions of *Yates v. the United States* (1957), *Communist Party v. Subversive Activities Committee Board* (1961), and *Albertson v. Subversive Activities Committee Board* (1965). During the 1960s the Court in general, and Justices Hugo Black and William Douglas in particular, were much more liberal—or literal—in their interpretation of the civil liberties contained in the Bill of Rights. Hamilton concurred with the thinking of Black and Douglas much more than the thinking of her old friend Justice Felix Frankfurter, with whom she regularly corresponded. As she stated to Charles Culp Burlingham in 1955, "I do believe in Jefferson's verdict that any doctrine should be permitted expression, provided truth is permitted to reply."[5]

The interesting triangular relationship between Hamilton, Frankfurter, and Burlingham deserves explanation. Frankfurter was arguably the most influential jurist of the twentieth century, serving as a law school professor at Harvard, major adviser to President Franklin Roosevelt, and a Supreme Court justice. Burlingham was a prominent lawyer and civic legal reform leader in New York City; many called him the "first citizen of New York." Hamilton was a close friend with and a frequent correspondent of both men. Hamilton and Frankfurter served sixteen years together on the Harvard faculty but had known each other well even before then, as both were charter members (1912) of Kitty Ludington's famous house parties in Lyme, Connecticut (see chapter 3).[6]

The three had in common their passion for liberal political reform, but Hamilton's relationship with Frankfurter cooled after he joined the Supreme Court in 1939 and practiced his famous philosophy of "judicial restraint." Frankfurter did not change his personal political thinking, but he believed that the high court should be very slow to reverse the decisions of the state and federal legislators. Hamilton was unhappy with him and expressed her dismay directly and forthrightly, especially in letters. She thought that the Frankfurter philosophy reduced the ability of the Court to serve as a needed check on lower-level reactionary—and even unconstitutional—decisions. In one of her letters she concluded with the declaration, "I have C.C. Burlingham with me." Usually she did, and Frankfurter acknowledged as much in a letter to Burlingham, referring to "your Alice—she was mine long ago."[7]

The "long-ago" period included Alice Hamilton's early years at Harvard, when she actively joined Frankfurter in the cause célèbre of defending and saving the immigrant radicals, Nicolas Sacco and Bartholemeo Vanzetti, from what they believed to be a miscarriage of justice. A daring 1920 daytime payroll robbery in a Boston suburb resulted in two gun deaths. At their 1921 trial, the two Italian laborers were convicted and sentenced to death in the midst of the post–World War I red scare. Legal appeals caused a delay of the execution until 1927. In the meantime, the case became tied to a larger issue, namely this question: Had America in the turn of the century admitted too many Southern and Eastern European workers who did not sufficiently understand and appreciate such traditional values as democracy and Protestantism, or was the opposition to the new immigration more nearly a case of elitism and even blatant prejudice and xenophobia? The liberals tended to assume that Sacco and Vanzetti were innocent, while conservatives tended to assume the opposite. Frankfurter wrote a lengthy defense of the convicts, and became the most publicly prominent person associated with their cause. Shortly before the execution, Hamilton was part of a delegation of six that met with Massachusetts governor Alvan Fuller in a vain effort to obtain a commutation of the death sentence.[8]

Throughout her life Hamilton sought to facilitate opportunity and justice for immigrants. Their babies were her medical clients at Hull House. Their breadwinners were her concern in the factories and mines. She protested efforts to deport aliens or to imprison them without due process. In her tenth decade of life, she wrote a stinging rebuke to Attorney General Nicholas Katzenbach for an action of the Justice Department against the American Committee for the Protection of the Foreign Born. Referencing the McCarthy era of the previous decade, she expressed a fear that "a second such era has begun."[9]

In seeking a summary interpretation of Alice Hamilton's old age, we find it to be the same as that of her preceding periods, namely that she never stopped learning, and she never stopped advocating based on that learning. Hamilton's most influential teacher was her mother, and the two most important lessons that her mother taught her were (1) "personal liberty was the most precious thing in life," and (2) "whatever was wrong in society was a personal concern for her and for us."[10] As the youthful Alice matured into adulthood, she learned to develop and refine the freedom that her mother had bestowed on her. In turn she learned how to bestow that freedom on others as she worked to empower both the poor, working-class immigrants to realize improved health and working conditions and society in general to be freed from the travails of war and restrictions of belief and expression. Hamilton herself learned how to be free to be a scientist and social scientist both, to investigate and uncover social problems, and to advocate for their elimination. She acquired the confidence that came with freedom so that when advocating for the freedom of expression of others—including even Nazis and Communists—she was not even intimidated by the FBI.[11] She learned to place great confidence in education and dialogue to resolve conflict peacefully and achieve reform.

If Hamilton's interpreters wished to write an epitaph for her memorial, they might choose the following: "Hamilton was free to do good."

NINE

—⟋⟋—

A PHOTOGRAPHIC MEMOIR

FAMILY AND YOUTH

The major repositories providing pictures for this chapter are the Allen County-Fort Wayne Historical Society; the Schlesinger Library at the Radcliffe Institute, Harvard University; the Bentley Historical Library of the University of Michigan; and the Jane Addams Hull House Museum Collection of the University of Illinois Chicago.

Alice's grandfather, Allen Hamilton (fig. 9.1), and grandmother, Emerine Holman Hamilton (fig. 9.2), were business, civic, and cultural leaders of Fort Wayne during the community's transformation from frontier town to postfrontier city. These portraits hang in the first-floor gallery of the Fort Wayne History Center Museum. The Hamilton Mansion (fig. 9.3) on the Hamilton family estate at the southeast edge of the city was one of the most elegant homes in the region and housed the city's finest library. It was located on the present site of the Anthis Career Center of the Fort Wayne Community Schools.

Her father, Montgomery Hamilton, appears in his Civil War uniform (fig. 9.4), and her mother, Gertrude Pond Hamilton, poses in what is also a young adult portrait (fig. 9.5). The family worshiped in the First Presbyterian Church (fig. 9.6), the congregation that the grandparents helped to found. The pictured structure, located several blocks north of the family estate, was built when Alice was a teenager.

Alice and her three sisters (fig. 9.7) were close in age and relationship, and the four sisters were also close to their many first cousins, with whom they played and studied on the estate. Alice was especially influenced by her older sister Edith Hamilton (fig. 9.8) and her cousins Agnes Hamilton (fig. 9.9) and Allen Hamilton Williams, the other members of "the three As" (fig. 9.10).

Figure 9.1. Alice's grandfather, Allen Hamilton. Courtesy of the Allen County-Fort Wayne Historical Society.

Figure 9.2. Alice's grandmother, Emerine Holman Hamilton. Courtesy of the Allen County-Fort Wayne Historical Society.

Figure 9.3. The Hamilton Mansion ("the big house"). Courtesy of the Allen County-Fort Wayne Historical Society.

Figure 9.4. Alice's father, Montgomery Hamilton. Courtesy of the Schlesinger Library at the Radcliffe Institute, Harvard University.

Figure 9.5. Alice's mother, Gertrude Pond Hamilton. Courtesy of the Schlesinger Library at the Radcliffe Institute, Harvard University.

Figure 9.6. Fort Wayne First Presbyterian Church. Courtesy of the Allen County-Fort Wayne Historical Society.

Figure 9.7. The four sisters: Norah, Margaret, Alice, and Edith (*left to right*). Courtesy of the Schlesinger Library at the Radcliffe Institute, Harvard University.

Figure 9.8. Edith Hamilton's graduation from Bryn Mahr College. Courtesy of the Schlesinger Library at the Radcliffe Institute, Harvard University.

Figure 9.9. Agnes and Alice Hamilton (left to right). Courtesy of the Schlesinger Library at the Radcliffe Institute, Harvard University.

Figure 9.10. The "Three As": Agnes Hamilton, Allen Williams, and Alice Hamilton (left to right). Courtesy of the Schlesinger Library at the Radcliffe Institute, Harvard University.

ALICE PORTRAITS BY AGE PERIOD

The series of Alice Hamilton portraits (figures 9.11–9.19) demonstrate the evolving physical appearance of Alice as she aged.

Figure 9.11. Alice as a young child. Courtesy of the Schlesinger Library at the Radcliffe Institute, Harvard University.

Figure 9.12. Alice and baby. Courtesy of the Schlesinger Library at the Radcliffe Institute, Harvard University.

Figure 9.13. Alice as an adolescent. Courtesy of the Schlesinger Library at the Radcliffe Institute, Harvard University.

Figure 9.14. Alice as a young adult. Courtesy of the Schlesinger Library at the Radcliffe Institute, Harvard University.

Figure 9.15. Alice in her twenties or thirties. Courtesy of the Schlesinger Library at the Radcliffe Institute, Harvard University.

Figure 9.16. Alice writing a report or letter, perhaps from Hull House. Courtesy of the Schlesinger Library at the Radcliffe Institute, Harvard University.

Figure 9.17. Meditative Alice, perhaps in her forties. Courtesy of the Schlesinger Library at the Radcliffe Institute, Harvard University.

Figure 9.18. Alice in her forties or fifties. Courtesy of the Schlesinger Library at the Radcliffe Institute, Harvard University.

Figure 9.19.   Alice late in her career as a physician and public health promoter. Courtesy of the National Institutes of Health.

MEDICAL SCHOOL

Alice completed her first year of medical studies (1890–91) at Taylor University, across the city from her home. The main campus building (fig. 9.20) is where she took her classroom work. This structure shows the original campus building, dating to the 1840s, with the 1884 addition joined on the front side. She completed her medical studies at the University of Michigan, graduating in 1893. She did her work there in the "Old Medical School Building" (fig. 9.21). Alice figures prominently in Horace Davenport's history of the Michigan Medical School, which includes a photograph of her in a physiology laboratory experiment with a male student (fig. 9.22). Her senior picture, by an Ann Arbor photographer, is one of her more distinguished portraits (fig. 9.23).

Figure 9.20. Taylor University in Fort Wayne, late nineteenth century. Courtesy of the Allen County-Fort Wayne Historical Society.

Figure 9.21. University of Michigan Medical School Building, late nineteenth century. Courtesy of Bentley Historical Library of the University of Michigan.

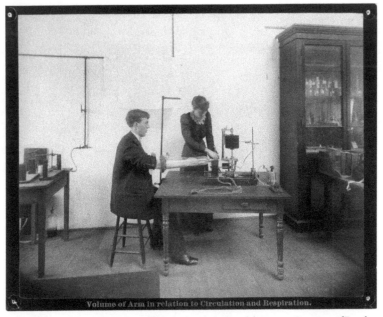

Volume of Arm in relation to Circulation and Respiration.

Figure 9.22. Alice in a University of Michigan physiology laboratory. Courtesy of Bentley Historical Library of the University of Michigan.

Figure 9.23. Senior portrait, Michigan Medical School, 1893. Courtesy of Bentley Historical Library of the University of Michigan.

HULL HOUSE RESIDENCY

In 1889 Jane Addams and Ellen Gates founded America's leading turn-of-the-century social settlement house to serve the large "Third Wave" (1880s–1910s) of American immigrants, who located primarily in industrial urban areas. Alice resided at Hull House for over two decades, 1897–1919, where she worked primarily as a physician and public health promoter. Figure 9.24 shows the exterior of Hull House, while figure 9.25 displays an inner courtyard. Alice Hamilton appears giving a speech with Jane Addams next to her in figure 9.26. Two early and long-serving leaders of Hull House, both of whom had a major impact on Alice, were Florence Kelley (fig. 9.27) and Julia Lathrop (fig. 9.28). Jane Addams was a national and international leader, and Alice worked and traveled with her both as her personal physician and also in promotion of their causes, including women's rights, international peace, and humanitarian relief. Figure 9.29 features Jane and Alice with a group of European women who were engaged in post–World War I hunger relief work. Figure 9.30 shows Alice aboard a ship in her international travel.

Figure 9.24. Hull House Complex (ca. 1900). CPC 01 A 0063 004, University of Illinois at Chicago Library, Special Collections.

Figure 9.25. Hull House Courtyard. JAMC 0000 0139 2657, University of Illinois at Chicago Library, Special Collections.

Figure 9.26. Alice standing and speaking with Jane Addams, seated to her left. JAMC 0000 0258 0512, University of Illinois at Chicago Library, Special Collections.

Figure 9.27. Florence Kelley. JAMC 0000 0262 0404, University of Illinois at Chicago Library, Special Collections.

Figure 9.28. Julia Lathrop. JAMC 0000 0267 0476, University of Illinois at Chicago Library, Special Collections.

Figure 9.29. Alice and Jane Addams in Europe, 1919 (post–World War I Relief Committee with Dr. Aletta Jacobs of the Netherlands, and Marion Fox and Joan Fry of Great Britain). Courtesy of the Schlesinger Library at the Radcliffe Institute, Harvard University.

Figure 9.30. Alice studying the ocean from a ship railing during one of her Atlantic voyages. Courtesy of the Schlesinger Library at the Radcliffe Institute, Harvard University.

THE HARVARD YEARS

Alice Hamilton spent most of her Harvard years (1919–35) in the new Harvard School of Public Health (HSPH) (fig. 9.31), which was organized in 1922 as an entity separate from the Harvard-MIT School of Health Officers but was still affiliated with the Harvard Medical School (fig. 9.32) until 1946. The officers who hired Alice in 1919 were Dean David Edsall (fig. 9.33) and President A. Lawrence Lowell (fig. 9.34).

Major colleagues with Alice in the early years of HSPH were the Drinker brothers, Cecil (fig. 9.35) and Philip (fig. 9.36). Cecil was a physician and physiologist and Philip an engineer and industrial hygienist. Felix Frankfurter (fig. 9.37) of the law school was one of Alice's closest friends on the faculty. He became an adviser of President Franklin Roosevelt and later a highly influential Supreme Court justice. Harriet Hardy (fig. 9.38) was one of Alice's most famous disciples. Ernest (fig. 9.39) and Katherine Codman were Alice's landlords and close friends. Their house on Beacon Hill (fig. 9.40) in Boston was also Alice's home during her sixteen years at Harvard.

Figure 9.31. Harvard School of Public Health. Courtesy of the Schlesinger Library at the Radcliffe Institute, Harvard University.

Figure 9.32. Harvard Medical School. Courtesy of the Schlesinger Library at the Radcliffe Institute, Harvard University.

Figure 9.33. Louisa Richardson Edsall, *David Linn Edsall* (Dean, Harvard Medical School), oil on canvas, 60.3 × 49.8 cm. Courtesy of the Harvard Portrait Collection, Harvard Medical School.

Figure 9.34. Harvard President A. Lawrence Lowell. Courtesy of the Schlesinger Library at the Radcliffe Institute, Harvard University.

Figure 9.35. Cecil Drinker, Harvard public health colleague. Courtesy of Louis Bachrach.

Figure 9.36. Philip Drinker, Harvard public health colleague. Courtesy of the Boston Children's Hospital Archives.

Figure 9.37. Felix Frankfurter, Harvard Law School colleague. Courtesy of the Library of Congress.

Figure 9.38. Irving Selikoff presenting the Alice Hamilton Award to Harriet Hardy, co-author with Hamilton. Courtesy of the Arthur H. Aufses, Jr., MD Archives, Icahn School of Medicine at Mount Sinai.

Figure 9.39. Ernest Amory Codman, Boston. Prominent surgeon and scientist and Hamilton landlord and friend. Courtesy of the Archives of the American College of Surgeons.

Figure 9.40. The Codman House on Beacon Hill. Courtesy of Joseph D. Brain.

RETIREMENT IN CONNECTICUT

When it came to planning for retirement living, Alice was exceptional in how early and thoughtfully she prepared. Spurred by the loss of the Fort Wayne family home to the Fort Wayne Public School System and especially the strong desire to keep the family together in one place, she enlisted the help of her longtime friend, dating to their academy days, Katherine (Kitty) Ludington, to find a comfortable and spacious house on the Connecticut River. There she spent summers beginning already in the Hull House period and then permanently after her Harvard retirement in 1935. The residents of the Hadlyme Home (fig. 9.41) included sisters Margaret and Norah, family friend Clara Landsberg, and Alice, while several cousins established a second family retirement home nearby. Figure 9.42 features a family reunion in the yard next to the river.

Figure 9.41. The Hadlyme Home from the Connecticut River. JAMC 0000 0258 3886, University of Illinois at Chicago Library, Special Collections.

Figure 9.42. Family picnic in the Hadlyme yard. Courtesy of the Schlesinger Library at the Radcliffe Institute, Harvard University.

HONORING ALICE

As a part of its millennium celebration in 2000, Alice's native city of Fort Wayne retained Hillsdale College sculptor Tony Frudakis to create a commemorative statue of Alice as part of the Hamilton Women Plaza in the newly designed Headwaters Park in downtown Fort Wayne (fig. 9.43). This memorial is only one of many awards and recognitions bestowed on Alice Hamilton throughout her life. For a comprehensive list of these honors, see table 9.1.

Figure 9.43. Alice Hamilton statue, nearing completion with sculptor Tony Frudakis, now standing in Headwaters Park, Fort Wayne, Indiana. Courtesy of the Allen County-Fort Wayne Historical Society.

Table 9.1. Honors and Awards

Organization	Recognition	Date
1. Bryn Mawr College	Master of Arts Degree	1896
2. United States Department of Labor	Certificate for Meritorious Service	1919
3. Mount Holyoke College	Honorary Degree	1925
4. American Association of University Women	Named Fellowship	1927
5. Smith College	Honorary Degree	1927
6. University Women of Chi Alpha Fraternity	National Achievement Award	1935
7. Tulane University	Honorary Degree	1937
8. American Association of Industrial Physicians and Surgeons	Honorary Membership	1938
9. University of Rochester	Honorary Degree	1938
10. American Industrial Hygiene Association	Cummings Award	1947
11. Lasker Foundation	Lasker Award	1947
12. Hobart and Smith College	Blackwell Centennial Citation	1949
13. National Consumers League	Fiftieth Anniversary Awards	1949
14. American Public Health Association	Forty-Year Membership	1950
15. Woman's Medical College of Pennsylvania	Honorary Degree	1950
16. Industrial Medical Association	Knudsen Awards	1953
17. New York Infirmary	Elizabeth Blackwell Citation	1953
18. American Industrial Hygiene Association	Honorary Membership	1955
19. *Time* Magazine	Woman of the Year in Medicine	1956
20. American Medical Women's Association	Medical Woman of the Year	1957
21. American Board of Industrial Hygiene	Toxicology Recognition	1962
22. Connecticut College	Hamilton Residence House Naming	1962
23. Hull House Association	Jane Addams Medal	1965
24. National Women's Hall of Fame	Induction	1973
25. American Chemical Society	Work Honored as a Chemical Landmark	1987

(continued)

Table 9.1. (continued)

26. Miss Porter's School	Hamilton Residence Hall Naming	c. 1985
27. National Institute for Occupational Safety and Health	Research Facility Naming	1987
28. National Institute for Occupational Safety and Health	Named Awards	1988
29. Connecticut Women's Hall of Fame	Induction	1994
30. U.S. Postal Service	Commemorative Stamp	1995
31. Harvard School of Public Health	Named Lectureship and Award	2011
32. University of Michigan Medical School	Alice Hamilton House Naming	2015

ENDNOTES

BRIEF EDUCATIONAL BIOGRAPHY

1. Alice Hamilton, *Exploring the Dangerous Trades: The Autobiography of Alice Hamilton* (Boston: Little, Brown, and Company, 1943), 18ff.; Barbara Sicherman, *Alice Hamilton: A Life in Letters* (Urbana: University of Illinois Press, 2003), 11–23.

1. PROLOGUE: ALICE HAMILTON ARRIVES AT HARVARD

1. "First Women," *Harvard Public Health Magazine* (Fall 2013); "Woman in Harvard Post: Dr. Alice Hamilton First of her Sex Elected to the Faculty," *New York Times*, March 12, 1919, 6; Letter, President and Fellows of Harvard College to Alice Hamilton, March 10, 1919, Archives of the Countway Library for the History of Medicine, Harvard University.

2. Catalogue, School for Health Officers, Harvard University and Massachusetts Institute of Technology, 1913–14 (Boston, 1913), 1.

3. Jean Alonzo Curran, *Founders of the Harvard School of Public Health, 1909–1946* (New York: Joseph Macy Jr. Foundation, 1970), 1–21.

4. Curran, *Founders of the Harvard School*, 17–18.

5. Curran, *Founders of the Harvard School*, 17–18; Hamilton, *Dangerous Trades*, 252–54; Sicherman, *Alice Hamilton*, 209–10, 217, 237; Letter, Alice Hamilton to Edith Hamilton, January 1919, in Sicherman, *Alice Hamilton*, 217–18. Catalogue, School for Health Officers, Harvard

University and Massachusetts Institute of Technology, 1916–17 (Boston, 1916), 12–13.

6. Jean Curran, interview with Alice Hamilton, November 29, 1963, Countway Library Archives, Harvard University.

7. Curran, *Founders of the Harvard School*, 18, 27–54; Sicherman, *Alice Hamilton*, 377, 397, 400; Harriet L. Hardy, "Beryllium Poisoning—Lessons in Control of Man-Made Disease," *New England Journal of Medicine* (November 25, 1965): 1188–99 (also delivered as the Alice Hamilton Lecture, Harvard School of Public Health, May 27, 1965); Lawrence K. Altman, "Dr. Harriet Hardy, Harvard Professor, Dies at 87," *New York Times*, October 15, 1993, B 00010; "Celebrating the Legacy of Thailand's 'Father of Public Health and Modern Medicine,'" news release, T.H. Chan School of Public Heath, Harvard University, 2018.

8. Curran, interview with Alice Hamilton.

9. Alice Hamilton, *Industrial Poisons in the United States* (New York: Macmillan, 1925); Alice Hamilton, *Industrial Toxicology* (New York: Harper and Brothers, 1934); Alice Hamilton and Harriett Hardy, *Industrial Toxicology*, 2nd ed. (Acton, MA, 1949); Raymond D. Harbison, Marie M. Bourgeois, and Giffe T. Johnson, eds., *Hamilton and Hardy's Industrial Toxicology*, 6th ed. (Hoboken, NJ: John Wiley and Sons, 2015); Mary Elizabeth Fouse Peyton, interview with Harriet Hardy, in *Some Pioneers in Industrial Hygiene*, ed. Charles D. Yaffe (Cincinnati: Annals of the American Conference of Governmental Industrial Hygienists, 1984), 7:76.

10. Alice Hamilton Faculty Record File Card, Harvard School of Public Health.

2. EARLY INFORMAL EDUCATION

1. Hugh McCulloch, *Men and Measures of Half a Century* (New York: Charles Scribner's Sons, 1889), 101–3.

2. Allyn C. Wetmore, "Allen Hamilton: The Education of a Frontier Capitalist" (PhD dissertation, Ball State University, Department of History, 1974), 154–56, 314, 319–20; Sicherman, *Alice Hamilton*, 14, 16–17; Hamilton, *Dangerous Trades*, 21–23; Albert Diserens, "Historic Sites," *Old Fort News* 20, nos. 3–4 (1957): 24–25; letter, Tom Castaldi to William Ringenberg, May 28, 2015; Charles R. Poinsatte, *Fort Wayne*

During the Canal Era, 1825–1855 (n.p., Indiana Historical Bureau, 1969), 9, 64, 95–96; John D. Beatty, ed., *History of Fort Wayne and Allen County* (Evansville: M. T. Publishing, 2006) 1:28–29, 43; Mina J. Carson, "Agnes Hamilton of Fort Wayne: The Education of a Christian Settlement House Worker," *Indiana Magazine of History*, vol. 80 (1984), 3; B. J. Griswold, *Pictorial History of Fort Wayne* (Chicago: Robert O. Law, 1917), 1:326, 422–23; James Madison, *Hoosiers: A New History of Indiana* (Bloomington: Indiana University Press, 2014), 87, 94, 120, 123; Mc-Culloch, *Men and Measures*, 105; Allen Hamilton Certificate of Naturalization, February, 1824, Wayne County (Indiana) Circuit Court, Allen Hamilton Papers, Indiana Historical Society; Last Will and Testament of Allen Hamilton, August 15, 1864, Allen Hamilton Papers, Indiana Historical Society. Also see Paul W. Gates, "The Role of the Land Speculator in Western Development," in Paul W. Gates, *The Jeffersonian Dream: Studies in the History of American Land Policy and Development* (Albuquerque: University of New Mexico Press, 1996), 6–22.

    3. Lorraine H. Davis and George R. Mather, "Emerine Jane Holman Hamilton," *Old Fort News* 63, no. 1 (2000): 1–3, 6; Hamilton, *Dangerous Trades*, 23–24; Tom Castaldi, "Allen County's Amazing Library," *Fort Wayne Monthly*, March 2011, http://historycenterfw.blogspot.com/2015/11; Sicherman, *Alice Hamilton*, 15; Holman Hamilton, "An Indiana College Boy in 1836: The Diary of Richard Henry Holman," *Indiana Magazine of History* 49, no. 3 (1953): 281; Israel George Blake, *The Holmans of Veraestau* (Oxford, OH: Mississippi Valley, 1943), xi; Paul W. Gates, *History of Public Land Law Development* (Washington, DC: U.S. Government Printing Office, 1968), 482.

    4. Hamilton, *Dangerous Trades*, 23–24.

    5. Tom Castaldi, "The Extraordinary Hamilton Family," historycenterfw.blogspot.com; Wetmore, "Allen Hamilton," 322–23; Hamilton, *Dangerous Trades*, 30–31, 41–42, 425–27; Sicherman, *Alice Hamilton*, 12–13, 380; Barbara Sicherman, *Well-Read Lives: How Books Inspired a Generation of American Women* (Chapel Hill: University of North Carolina Press, 2010), 80–85; Beatty, *Fort Wayne*, 222–23.

    6. Sicherman, *Well-Read Lives*, 90, 101, 107; Carson, "Agnes Hamilton," 15–18; Griswold, *Fort Wayne*, 446; www.geni.com/people/montgomery-hamilton.

3. LEARNING IN TRANSITION TO ADULTHOOD

1. During their "growing up" period in Fort Wayne, it was first cousins Agnes, Alice, and Allen Williams—"the three As," as they called themselves—who were especially close in age, location, affection, and playing habits. Alice noted that Edith "ran with" the older cousins, and she with the younger group. Allen, like Alice, became a doctor, earning his degrees at Harvard and marrying another physician, the Boston-born Marian Bartholow Walker. Agnes Hamilton, diary, July 24, 1887; and Sicherman, *Alice Hamilton*, 13, 100.

2. Hamilton, *Dangerous Trades*, 18–19, 32, 53–56; Judith P. Hallett, "Edith Hamilton," in Garraty and Carnes, eds. *American National Biography*, 9:918–20; John Mason Brown, "The Heritage of Edith Hamilton, 1867–1963," *Saturday Review*, June 22, 1963, 17.

3. Alice stated that "the big house" was "as important a background for our lives as our own place." Davies and Mather, *Emerine Hamilton*, 11.

4. Castaldi, "Hamilton Family"; Diserens, "Historic Sites," 24–25; Griswold, *Fort Wayne*, 411, 468; Bill Griggs and Jim Nitz, "Kekionga Ball Grounds," Society for American Baseball Research, http://sabr.org; Sicherman, *Alice Hamilton*, 22, 197, 237, 245, 349, 376; Hamilton, *Dangerous Trades*, 405–7; Davis and Mather, "Emerine Hamilton," 11; Doris Fielding Reid, *Edith Hamilton: An Intimate Portrait* (New York: W. W. Norton and Company, 1967), chapter 4; Carson, "Agnes Hamilton," 32; Peyton, "Harriet Hardy," 79.

5. "Veraestau," indianalandmarks.org; Davis and Mather, "Emerine Hamilton," *Old Fort News* 63, no. 2000, 8, 11; *The Holmans of Veraestau*, 7–8; Hamilton, *Dangerous Trades*, 29, 35–37; Sicherman, *Alice Hamilton*, 22–23, 195–96; Jim Lampos and Michelle Pearson, "Looking Back: Legacy of Katherine Ludington and That of Her Grand Home," *Shoreline Times*, August 2, 2015; Agnes Hamilton, diary, June 18, 1887; William C. Ringenberg, *The Christian College: A History of Protestant Higher Education in America*, 2nd ed. (Grand Rapids, MI: Baker Academic, 2006), 67–68; Nancy Davis and Barbara Donahue, *Miss Porter's School: A History* (Farmington, CT: Miss Porter's School, 1992), 18, 103; Eugenia Peretz, "The Code of Miss Porter's, *Vanity Fair*, July, 2009, www.vanityfair.com; Carson, "Agnes Hamilton," 8; Hamilton, *Dangerous Trades*, 270–72; letter, Alice Hamilton to Clara Landsberg in

Sicherman, *Alice Hamilton*, 282; Andrew L. Warshaw, "Presidential Address: Achieving our Personal Best," *Bulletin of the American College of Surgeons*, December 1, 2014.

6. The experience at Miss Porter's School for Young Ladies did not encourage Alice toward singleness. On the contrary, Sarah Porter's goal was to educate young women to be "good Christians, good wives and companions to their husbands, and good mothers." Davis and Donahue, *Miss Porter's School*, 7.

7. Sicherman, *Alice Hamilton*, 17–21, 31–32, 88–89; Hamilton, *Dangerous Trades*, 31–32, 38; Sicherman, *Well-Read Lives*, 98, 107; also see letter, Alice Hamilton to Agnes Hamilton, December 6, 1896, in Sicherman, *Alice Hamilton*, 105–6.

8. Note, for example, Edith's books on biblical interpretation, including *Spokesmen for God*, on the Old Testament prophets; and *Witness to Truth: Christ and His Interpreters*. The Second Great Awakening (1800–1835), led by Yale professor Nathaniel Taylor in ideas and Oberlin College president Charles G. Finney in action, represented America's national conversion to Free Will Theology. The Presbyterian Church nationally, and in Fort Wayne in particular, was split into "old school" and "new school" (pro-revival, usually antislavery) factions with the new school "heavy weights" (Lyman Beecher and sons Henry Ward and Edward) coming to town to try to convert the Hamilton family church to their views. They were unsuccessful. Montgomery Hamilton was a small child at the height of this controversy, and it undoubtedly influenced his later strong interest in theology in general and supporting the traditional Presbyterianism in particular. Poinsatte, *Fort Wayne*, 151–60.

9. Hamilton, *Dangerous Trades*, 26–29; Doris Fielding Reid, *Edith Hamilton: An Intimate Portrait* (New York: W. W. Norton and Company, 1967), 23, 30–31; Edith Hamilton, *Witness to the Truth: Christ and His Interpreters* (New York: W. W. Norton and Company, 1948), 204–12; geni.com/montgomery-hamilton; Sicherman, *Alice Hamilton*, 56, 89; letter, Alice Hamilton to Agnes Hamilton, October 13, 1919, in Sicherman, *Alice Hamilton*, 246; Horace W. Davenport, *Not Just Any Medical School: The Science, Practice, and Teaching of Medicine at the University of Michigan, 1850–1941* (Ann Arbor: University of Michigan Press, 1999), 336.

10. Ever since their "Plan of Union of 1801," the Presbyterians and Congregationalists, in their efforts of western expansion, had worked closely together.

11. Sicherman, *Alice Hamilton*, 14–19; Tom Castaldi, "Hamilton Family," 40–41; Fort Wayne Female College, Trustee Minutes, March 15, 1849, and April 22, 1851; A. H. Hamilton, Obituary, *Fort Wayne Gazette*, May 10, 1895, 1, familysearch.org; "Andrew Holman Hamilton," *Biographical Directory of the American Congress*, bioguide.congress.gov; Diserens, "Historic Sites," 24–25; Taylor University, *1891–92 Catalog*, 4; Fort Wayne College of Medicine, *1890 Catalog*, 3, and *1891 Catalog*, 1.

4. MEDICAL SCHOOLS

1. Hamilton, *Dangerous Trades*, 38; William C. Ringenberg, *Taylor University: The First 150 Years* (Grand Rapids, MI: Eerdmans, 1996), 22–23; Sicherman, *Alice Hamilton*, 18, 23.

2. Hamilton, *Dangerous Trades*, 38; Fort Wayne College of Medicine, *1890 Catalog*, 3; Ringenberg, *Taylor University*, 62–64.

3. Professor George W. McCaskey began a medical journal, *McCaskey's Clinical Studies*, in 1878. It continued under changed names throughout the nearly three-decade history of FWMC. When in 1892 ophthalmologist Albert E. Bulson joined the faculty, he soon assumed the editorship of what was then called the *Fort Wayne Medical Magazine*. Then, in the same year that FWMC became a part of the quite new (founded 1903) Indiana University School of Medicine in 1908, the state medical association—after having discussed the need for a statewide journal for two years or so—recruited the former FWMC faculty member and journal editor Bulson to become the founding editor of its new publication, the *Journal of the Indiana State Medical Association*. Edward L. Van Bushkirk, "Fort Wayne Medical Schools," *Indiana Medical History Quarterly* (March 1977): 9; "History," About the ISMA, Indiana State Medical Association, ismanet.org; editorial, "The Journal of the Indiana State Medical Association," *The Journal of the Indiana State Medical Association*, January 15, 1908, 18.

4. Sicherman, *Alice Hamilton*, 18, 23, 31–35, 40–41; Hamilton, *Dangerous Trades*, 29–30, 36, 38; *Fort Wayne College of Medicine Catalog*, 1885–86, 2–7; *Taylor University and Fort Wayne College of Medicine*

Catalog, 1890, 2, 4, 17, 24–25; Fort Wayne College of Medicine Catalog, *1891*, 1, 3–4, 8–9, 12; *Taylor University and Fort Wayne College of Medicine Catalog*, 1891–92, 7; *Fort Wayne College of Medicine Catalog*, 1892–3, 2–3; *Fort Wayne College of Medicine Catalog*, 1893–94, 2–5; "Historic Buildings and Structures of the West Central Neighborhood Association," Fort Wayne, Indiana, www.westcentralneighborhood.org; E. Steryl Phinney, "Christian N. Stemen," *Taylor Bulletin*, September 1962, 6–8; Dorothy Ritter Russo, ed., *One Hundred Years of Indiana Medicine*, 1849–1949 (n.p.: Indiana State Medical Association, 1949), 37, 66.

5. Hamilton, *Dangerous Trades*, 38–53; Madeline Grant, *Alice Hamilton: Pioneer Doctor in Industrial Medicine* (New York: Abelard-Schuman, 1967), 36–55; Sicherman, *Alice Hamilton*, 35–57, 63–64, 72–72, 76, 88–91, 104, 108; Kenneth M. Ludmerer, *Learning to Heal: The Development of American Medical Education* (New York: Basic Books, 1985), 14–15, 47, 56–61, 72–73, 166–176; Davenport, *Medicine at Michigan*, 31; A. McGehee Harvey, et al., *A Model of Its Kind: A Centennial History of Medicine at Johns Hopkins*, 2 vols. (Baltimore: John Hopkins University Press, 1989), 1:22–23, 27–28, 31, 38–39, 50; Martin Kaufman, "American Medical Education," 18, in Ronald L. Numbers, ed., *The Education of American Physicians* (Berkeley: University of California Press, 1980).

6. Hamilton, *Dangerous Trades*, 53–56, 60.

7. Louise Caroll Wade, "Settlement Houses," *Encyclopedia of Chicago* (Chicago: Chicago Historical Society, 2005), 1135; Allen F. Davis, *Spearheads for Reform: The Social Settlements and the Progressive Movement, 1890–1914* (New York: Oxford University Press, 1967), chap. 2, 6, 7, 10; Judith Ann Trolander, *Professionalism and Social Change: From the Settlement House Movement to Neighborhood Centers* (New York: Columbia University Press, 1987) 7–8, 183ff.

5. LEARNING SELF-CONFIDENCE AT HULL HOUSE

1. Letter, Alice Hamilton to Agnes Hamilton, March 6, 1892; Hamilton, *Dangerous Trades*, 47; Letter, Alice Hamilton to Agnes Hamilton, August 1891.

2. Agnes Hamilton, diary, August 18, 1889; Sicherman, *Alice Hamilton*, 15; Last Will and Testament of Emerline J. Hamilton, Allen County, Indiana, Court House.

3. Jean Bethke Elshtain interview, as referenced in Vicki J. McCoy, "Alice Hamilton: The Making of a Feminist-Pragmatist Rhetor" (MA thesis, Georgia State University, 2006), 35; Hamilton, *Dangerous Trades*, 16.

4. Hamilton, *Dangerous Trades*, 16, 61; Letters, Alice Hamilton to Agnes Hamilton, March 1898, August 9, 1898, and August 24, 1898.

5. Hamilton, *Dangerous Trades*, 63–64.

6. Hamilton, *Dangerous Trades*, 64–67.

7. Sicherman, *Alice Hamilton*, 9; Letter, Alice Hamilton to Agnes Hamilton, March 5, 1893; Catherine E. Forrest Weber, "A Citizen of Athens: Fort Wayne's Edith Hamilton," *Traces of Indiana and Midwestern History*, Volume 14, Number 1, Winter 2002; Janice Lee Jayes, "Hamilton, Edith (1867–1963)" in Anne Commire, ed. *Women in World History: A Biographical Encyclopedia* (Detroit: Yorkin, 2002) 728.

8. Mark H. Senter III, *When God Shows Up: a History of Protestant Youth Ministry in America*, (Grand Rapids, MI: Baker Academic, 2010), 151–168; Carson, "Agnes Hamilton," 7ff.

9. Letters, Alice Hamilton to Agnes Hamilton, February 12, 1888; December 11, 1892; July 23, 1893.

10. Agnes Hamilton, diary, March 9, 1884 and January 3, 1885, as found in Carson, "Agnes Hamilton," 4–5.

11. Sicherman, *Alice Hamilton*, 112, 130; Letter, Alice Hamilton to Agnes Hamilton, August 28, 1902 in Sicherman, *Alice Hamilton*, 129–30.

12. Alice Hamilton to Agnes Hamilton, July 3, 1898; Hamilton, *Dangerous Trades*, Page 79; Alice Hamilton to Agnes Hamilton, August 8, 1900; Hull House Bulletins, March 1898; January/February 1899; November/December 1899; Mid-winter 1903–1904; Autumn 1904; Hamilton, *Dangerous Trades*, 69–70; Alice Hamilton et al., "The Midwives of Chicago," *Journal of American Medical Association* 50 (April 25, 1908), 1346–50.

6. INVESTIGATING THE DANGEROUS TRADES

1. Hamilton, *Dangerous Trades*, 98; Alice Hamilton, "The Fly as a Carrier of Typhoid," *Journal of the American Medical Association* (February 28, 1903): 576–83.

2. Hamilton, "Fly as a Carrier of Typhoid," 576–83; Alice Hamilton, "The Chicago Water Supply," *Journal of the American Medical Association* 40, no. 11 (1903): 74.

3. Hamilton, *Dangerous Trades*, 100–3; *Hull House Bulletin* (Autumn 1904): 21; *Chicago Daily Tribune* (*Chicago Tribune*), "Cocaine Trade in Disguise," September 23, 1907; *Hull House Yearbook*, 1906–7, 61.

4. G. Leighton and C. Bargiel, "A History of Illinois Drug Control Laws from 1818 to 1975," *John Marshall Journal of Practice and Procedure* (1975): 148.

5. *Hull House Bulletin* (Autumn 1904): 21; "Jessie Binford Dies, Worked at Hull House," *Chicago Tribune*, July 11, 1966, section 3, p. 10; Hamilton, *Dangerous Trades*. 100–101; Jessie Binford, "Report of Work During March of West Wide Office of the Legal Aid Society at Hull House in Cocaine Cases," *Quarterly Review of Legal Aid Society of Chicago* 3, no. 1 (April 1906): 3–4; *Hull House Bulletin* (Autumn 1904): 21; *Hull House Yearbook*, 1906–7, 61; Jessie Binford, "Profit in Child Victims to Cocaine," *Charities and the Commons* 9 (September 1904): 423.

6. *Hull House Bulletin* (Autumn 1904): 21; Binford, "Profit in Child Victims to Cocaine," 423.

7. Hamilton, *Dangerous Trades*, 100–3; Binford, "Report of Work," 3–4.

8. *Hull House Yearbook*, 1906–7, 61; "Cocaine Trade in Disguise"; "A New Weapon Against Cocaine," *Charities and the Commons* (November 16, 1907), 1045; Leighton and Bargiel, "History of Illinois Drug Control Laws," *John Marshall Journal of Practice and Procedure*, 148; Hamilton, *Dangerous Trades*, 102.

9. Sicherman, *Alice Hamilton*, 153; Hamilton, *Dangerous Trades*, 114–15.

10. Maurine Weiner Greenwald and Margo J. Anderson, *Pittsburgh Surveyed: Social Science and Social Reform in the Early Twentieth Century* (Pittsburgh, PA: University of Pittsburgh Press, 1996).

11. Hamilton, *Dangerous Trades*, 117; Department of Commerce and Labor, *Bulletin of the Bureau of Labor*, no. 86 (January 1910): 21, 39–41, 67–85; letters with Jonathan Schoer, March 23, 2018.

12. Alice Hamilton, "Occupational Conditions of Tuberculosis," *Charities and the Commons* 16 (May 5, 1906): 205–7; Alice Hamilton, "Industrial Diseases. With Special Reference to the Trades in Which Women are Employed," *Charities and the Commons* 20 (September 5, 1908): 655–59.

13. Hamilton, *Dangerous Trades*, 118; Sicherman, *Hamilton*, 156.

14. John Commons was already nationally known for promoting a version of Georgism, in which each individual would personally own

the fruits of his or her labor, while wealth from land and other natural resources was to be owned by society more generally. He served as executive editor of two major histories of the labor movement in the United States, *A Documentary History of American Industrial Society* and *History of Labor in the United States*. His ideas formed the basis of Wisconsin's worker's compensation law, the first in the nation. See J. David Hoeveler Jr., "John R. Commons," *Historical Dictionary of the Progressive Era, 1890–1920*, rev. ed. (Westport, CT: Greenwood, 1988), 85–86.

15. John Andrews is credited as the AALL's most influential leader, having expanding its role into active promotion of various labor laws and regulations, including, but much broader than, safety and industrial poisons. In 1911 he founded and began his tenure as editor of the *American Labor Legislation Review*. He also contributed to John Commons's *History of Labor in the United States*. (See John D. Chasse, "The American Association for Labor Legislation: An Episode in Industrialist Policy Analysis," *Journal of Economic Issues* 25, no. 3 [September]: 800.)

16. Irene Osgood (Andrews) studied at the New York School of Philanthropy (now Columbia University School of Social Work) and the University of Wisconsin and interned at the Wisconsin University Settlement House in Milwaukee. Her professional efforts included international peace and immigration (she would briefly lead the Northwestern Settlement House in Chicago), but primarily women in the workplace. (See Irene Osgood Andrews and John W. Leonard, eds., *Women's Who's Who of America 1914–1915* [New York: American Commonwealth Company], 51.)

17. Hamilton, *Dangerous Trades*, 117.

18. John D. Chasse, "The American Association for Labor Legislation: An Episode in Industrialist Policy Analysis," *Journal of Economic Issues* 25, no. 3 (September 1991): 800; Sicherman, *Alice Hamilton*, 154; Hamilton, "Industrial Diseases: With Special Reference to the Trades in Which Women Are Employed," *Charities and the Commons* 20 (1908): 655–58; Hamilton, *Dangerous Trades*, 117; *Bulletin of the Bureau of Labor*, no. 86, 31; Alton Lee, "The Eradication of Phossy Jaw: A Unique Development of Federal Police Power," *The Historian* 29, no. 1 (November 1966): 6.

19. Sicherman, *Alice Hamilton*, 156–57; Hamilton, *Dangerous Trades*, 119; *Bulletin of the Bureau of Labor*, no. 86, 32; *Report of Commission on*

Industrial Diseases to His Excellency, Governor Charles S. Deneen, January 1911, 49–83, 88–97, 154.

20. Hamilton, *Dangerous Trades*, 120–21.

21. Hamilton, *Dangerous Trades*, 126–29.

22. Hamilton, *Dangerous Trades*, 128; "Women in the Lead Industries," *Bulletin of the United States Bureau of Labor Statistics*, no. 253 (1919); "The White Lead Industry in the United States, with an Appendix on the Lead-Oxide Industry," *Bulletin of the United States Bureau of Labor Statistics*, no. 95 (January 1911); "Lead Poisoning in Potteries, Tile Works, and the Porcelain Enameled Sanitary Ware Factories," *Bulletin of the United States Bureau of Labor Statistics,* no. 104 (January 1912); "Hygiene of the Painter's Trade," *Bulletin of the United States Bureau of Labor Statistics,* no. 120 (January 1913); "Lead Poisoning in the Smelting and Refining of Lead," *Bulletin of the United States Bureau of Labor Statistics,* no. 141 (January 1914); "Lead Poisoning in the Manufacture of Batteries," *Bulletin of the United States Bureau of Labor Statistics,* no. 165, (Department of Commerce and Labor. January 1915).

23. Hamilton, *Dangerous Trades*, 124; Alice Hamilton, "The Economic Importance of Lead Poisoning," *Bulletin of America* 40, no. 5 (October 1914): 299–304; *Report of Commission on Industrial Diseases to His Excellency, Governor Charles S. Deneen,* January 1911, 21–22; Alice Hamilton, "The White Lead Industry in the United States," 3rd International Congress for Trade Diseases, later printed in *Journal of the Austrian Medical Service,* special edition Sicknesses You Get on Your Job 1918): 5.

24. Hamilton, *Dangerous Trades*, 63, 131–37, 185–89, 201; Alice Hamilton, *Industrial Poisons in the United States* (New York: Macmillan, 1925), vi; Leslie Nickels, *Education of Alice Hamilton: Curriculum for Taking up the Cause of the Working Class,* 18, 20.

25. "Carbon-Monoxide Poisoning," *Bulletin of the United States Bureau of Labor Statistics,* no. 291 (January 1921); "Industrial Poisons Used in the Rubber Industry," *Bulletin of the United States Bureau of Labor Statistics,* no. 179 (January 1915); "Hygiene of the Printing Trades," *Bulletin of the United States Bureau of Labor Statistics,* no. 209 (January 1917); "Industrial Poisoning in Making Coal-Tar Dyes and Dye Intermediates," *Bulletin of the United States Bureau of Labor Statistics,* no. 280 (January 1921); "Occupational Poisoning in the Viscose Rayon Industry," *Bulletin of the United States Bureau of Labor Statistics,* no. 34 (1921).

26. Sicherman, *Hamilton*, 184–94; Hamilton, *Dangerous Trades*, 183; Alice Hamilton, "Industrial Poisons Encountered in the Manufacture of Explosives," *Journal of the American Medical Association* 68, no. 20 (May 19, 1917): 1445–49.

27. Hamilton, *Dangerous Trades*, 184, 189–91, 201, 206–7; Sicherman, *Alice Hamilton*, 200; Joseph Miller to Alice Hamilton, March 19, 1917; "Effect of the Air Hammer on the Hands of Stonecutters," *Bulletin of the Bureau of Labor*, No. 236 (January–July 1918): 5, 54–58.

28. "Report on the President's Mediation Commission to the President of the United States," January 9, 1918, 4, "Report on the Brisbee Deportations. Made by the President's Mediation Commission to the President of the United States," Department of Labor, November 6, 1917, 4–5; Hamilton, *Dangerous Trades*, 208–10.

29. Alice Hamilton Notes, New Cordelia Mine, Ajo, January 11, 1919; Alice Hamilton Notes, Old Dominion Mine, Globe, January 16, 1919; Alice Hamilton Notes, Inspiration Mine, Miami, January 17, 1919; Alice Hamilton Notes, Miami Copper Company Mine, January 18, 1919; Alice Hamilton Notes, Arizona Copper Mine, Copper Hill near Globe, January 20, 1919; Alice Hamilton Notes, Phelps-Dodge Corporation, Yankee Mine, January 23, 1919; Alice Hamilton Notes, Arizona Copper Company, Humboldt Mine, January 23, 1919.

30. Alice Hamilton Notes, Old Dominion Mine, Globe, January 16, 1919; Hamilton, *Dangerous Trades*, 210–18.

31. Hamilton, *Dangerous Trades*, 223–24, 254.

7. THE SCIENTIST AS SOCIAL SCIENTIST

1. Hamilton, *Dangerous Trades*, 83–91; letter, Alice Hamilton to Agnes Hamilton, August 9, 1898, Alice Hamilton Papers, Schlesinger Library, Radcliffe College, Harvard University.

2. Alice Hamilton, "Excessive Childbearing as a Factor in Infant Mortality," *Bulletin of the American Academy of Medicine* 11 (1910): 181–87.

3. Alice Hamilton, "Prostitutes and Tuberculosis," *The Survey*, February 17, 1917, 516–17; Alice Hamilton, "Rehabilitation of Boiled Milk," *The Survey*, December 13, 1913, 303.

4. Sicherman, *Alice Hamilton*, 148–49; Eleanor J. Stebner, *The Women of Hull House: Their Spirituality, Vocation, and Friendship*

(Albany: University of New York Press, 1997), 141; *Women of Hull House*, 139, refers to and supports Madeleine Grant, *Alice Hamilton: Pioneer Doctor in Industrial Medicine* (New York: Abelard, 1967); Letter, Alice Hamilton to Katy Bowditch Codman, January 13, 1959, Alice Hamilton Papers.

5. Letter, Alice Hamilton to Agnes Hamilton, April 5, 1915, in Sicherman, *Alice Hamilton*, 184–85; Letter, Alice Hamilton to Mary Rozet Smith, April 22, 1915, in Sicherman, *Alice Hamilton*, 185–86; Alice Hamilton, "As One Woman Sees the Issue," *The New Republic*, October 7, 1916, 239–41.

6. Hamilton, *Dangerous Trades*, 127, 164; Sicherman, *Alice Hamilton*, 185; Letter, Alice Hamilton to Mary Rozet Smith, May 5, 1915, in Sicherman, *Alice Hamilton*, 189–90; Jane Addams, Emily Green Balch, and Alice Hamilton, *Women at the Hague: The International Congress of Women and its Results* (New York: Macmillan, 1915).

7. Letter, Alice Hamilton to Margaret Hamilton, April 23, 1919, in Sicherman, *Alice Hamilton*, 233; Jane Addams and Alice Hamilton, "Official Report of Jane Addams and Dr. Alice Hamilton to the American Society of Friends Service Committee" (1919); Jane Addams and Alice Hamilton, "After the Lean Years," *The Survey*, September 24, 1919; Alice Hamilton, "On a German Railway Train," *The New Republic*, September 24, 1919, 234–35.

8. Alice Hamilton, "Because War Breeds War," chap. 8 in Rose Young, ed., *Why Wars Must Cease* (New York: Macmillan, 1935).

9. Hamilton, *Dangerous Trades*, 58, 80, 85; Sicherman, *Alice Hamilton*, 15.

10. Hamilton, *Dangerous Trades*, 76, 85–86, 101.

11. *Chicago Record-Herald*, March 7–8, 1911.

12. Hamilton, *Dangerous Trades*, 110.

8. EPILOGUE: THE SENIOR AS A PUBLIC INTELLECTUAL

1. Alice Hamilton Personnel Records, T.H. Chan School of Public Health, Harvard University; Hamilton, *Dangerous Trades*, 387–404; Sicherman, *Alice Hamilton*, 357–58, 372–73, 377; Peyton, "Harriet Hardy," 78.

2. Hamilton, *Dangerous Trades*, 395–98; Sicherman, *Alice Hamilton*, 378, 380–83.

3. Alice's reading material included the mainstream *New York Times, Washington Post,* and *Manchester Guardian* (from England); *The Nation* (which her father had enjoyed); *The New Republic* (which had been founded by her friends, Herbert Croly and Walter Lippman); the literary *Saturday Review;* the left-wing *The Worker;* and the right-wing *National Review.*

4. Hamilton, *Dangerous Trades,* 422, 425; Sicherman, *Alice Hamilton,* 380ff.

5. Sicherman, *Alice Hamilton,* chapters 9 and 10; letter, Alice Hamilton to Charles Culp Burlingham, March 1, 1955, in Sicherman, *Alice Hamilton,* 396–97; James M. Smith and Paul L. Murphy, eds., *Liberty and Justice,* 2 vols. (New York: Alfred A. Knopf, 1968), vol. 2, chapter 28, 13. Alice Hamilton file, Federal Bureau of Investigation, US Department of Justice.

6. H. N. Hirsch, *The Enigma of Felix Frankfurter* (New York: Basic Books, 1981), 3, 193; Sicherman, *Alice Hamilton,* 382–83, 386; Hamilton, *Dangerous Trades,* 405; Letter, Alice Hamilton to Charles Culp Burlingham, January 14, 1953, in Sicherman, *Alice Hamilton,* 386–88.

7. Sicherman, *Alice Hamilton,* 382–83, 386, 405; letter, Alice Hamilton to Felix Frankfurter, March 2, 1951, in Sicherman, *Alice Hamilton,* 403–404; Melvin I. Urofsky, "The Failure of Felix Frankfurter," *University of Richmond Law Review* 26, issue 1 (1991): 186–96.

8. Felix Frankfurter, *The Case of Sacco and Vanzetti* (Boston: Little, Brown, and Company, 1927), 3–110; Hamilton, *Dangerous Trades,* 273–79; Hirsch, *Enigma of Frankfurter,* 90–94; Harlan B. Phillips, *Felix Frankfurter Reminisces* (New York: Reynal and Company, 1960), 202–17.

9. Sicherman, *Alice Hamilton,* 244, 414; Alice Hamilton to Nicholas Katzenbach, July 15, 1965, Alice Hamilton file, Department of Justice, Washington, DC.

10. Hamilton, *Dangerous Trades,* 32.

11. Alice Hamilton file, Federal Bureau of Investigation, US Department of Justice, entries December 30, 1946; April 28, 1954; June 17, 1954; December 17, 1955; March 26, 1960; June 14, 1960; July 21, 1965; September 23, 1965; and February 3, 1966; Barry Castleman, "Alice Hamilton and the FBI," *International Journal of Occupational and Environmental Health* 22, no. 2 (April 2016): 173–74.

BIBLIOGRAPHY

1. Wilma R. Slaight has granted the authors permission to reprint the parts of her dissertation ("Alice Hamilton: First Lady of Industrial Medicine," Case Western Reserve University, 1974) bibliography that lists the writings of Alice Hamilton, even while encouraging us to find additional writings. In the few cases where we have found others, we have prefaced them with an asterisk. Letter, Wilma Straight to Joseph D. Brain, William C. Ringenberg, and Matthew C. Ringenberg, September 15, 2017.

BIBLIOGRAPHY

WILMA R. SLAIGHT BIBLIOGRAPHY OF THE WRITINGS OF ALICE HAMILTON[1]

Note: These publications and all sublists within this bibliography are listed chronologically rather than alphabetically.

1. GOVERNMENT PUBLICATIONS

"The White Lead Industry in the United States, With an Appendix on the Lead-Oxide Industry." *Bulletin of the U.S. Bureau of Labor*, no. 95. Washington, DC: Government Printing Office, 1911.

"Lead Poisoning in Potteries, Tile Works, and the Porcelain Enameled Sanitary Ware Factories." *Bulletin of the U.S. Bureau of Labor*, no. 104. Washington, DC: Government Printing Office, 1912.

"Hygiene of the Painter's Trade." *Bulletin of the U.S. Bureau of Labor Statistics*, no. 120. Washington, DC: Government Printing Office, 1913.

"Lead Poisoning in the Smelting and Refining of Lead." *Bulletin of the U.S. Bureau of Labor Statistics*, no. 141. Washington, DC: Government Printing Office. 1914.

"Lead Poisoning in the Manufacture of Storage Batteries." *Bulletin of the U.S. Labor Statistics*, no. 165. Washington, DC: Government Printing Office. 1915.

"Industrial Poisons Used in the Rubber Industry." *Bulletin of the U.S. Bureau of Labor Statistics*, no. 179. Washington, DC: Government Printing Office, 1915.

*With Charles H. Verrill. "Hygiene of the Printing Trades." *Bulletin of the U.S. Bureau of Labor Statistics*, no. 209. Washington, DC: Government Printing Office, 1917.

"Industrial Poisons Used or Produced in the Manufacture of Explosives." *Bulletin of the U.S. Bureau of Labor Statistics*, no. 219. Washington, DC: Government Printing Office, 1917.

"Effect of the Air Hammer on the Hands of Stonecutters." *Bulletin of the U.S. Bureau of Labor Statistics*, no. 236. Washington, DC: Government Printing Office, 1918.

"Women in the Lead Industries." *Bulletin of the U.S. Bureau of Labor Statistics*, no. 253. Washington, DC: Government Printing Office, 1919.

"Industrial Poisoning in Making Coal-Tar Dyes and Dye Intermediates." *Bulletin of the U.S. Bureau of Labor Statistics*, no. 280. Washington, DC: Government Printing Office, 1921.

"Carbon-Monoxide Poisoning." *Bulletin of the U.S. Bureau of Labor Statistics*, no. 291. Washington, DC: Government Printing Office, 1921.

"Women Workers and Industrial Poisons." *Bulletin of the Women's Bureau*, no. 57. Washington, DC: Government Printing Office, 1926.

"Recent Changes in the Painter's Trade." *Bulletin of the Department of Labor, Division of Labor Standards*, no. 7. Washington, DC: Government Printing Office, 1936.

"Occupational Poisoning in the Viscose Rayon Industry." *Bulletin of the U.S. Department of Labor Division of Labor Standards*, no. 34. Washington, DC: Government Printing Office, 1940.

2. BOOKS AND PAMPHLETS

With Jane Addams. *Official Report of Jane Addams and Dr. Alice Hamilton to the 'American Society of Friends' Service Committee, Philadelphia, on the Situation in Germany*. Nebraska Branch, American Relief Fund for Central Europe, n.d.

With Jane Addams. "The 'Piece-Work' System as a Factor in the Tuberculosis of Wage-Workers." In *Transactions of the Sixth International Congress on Tuberculosis*. Washington, DC, September 28–October 5, 1908.

"Report by Dr. Alice Hamilton on Investigations of the Lead Troubles in Illinois, from the Hygienic Standpoint." *Report of Commission on*

Occupational Diseases to His Excellency Governor Charles S. Deneen. Chicago: Warner Printing Co., 1911.

With Jane Addams and Emily G. Balch. *Women at The Hague: The International Congress of Women and its Results.* New York: Macmillan, 1915.

"Occupational Diseases in Illinois." *Report of the Health Insurance Commission of the State of Illinois.* Springfield: Illinois State Journal Co., 1919.

Poverty and Birth Control. New York: American Birth Control League, 1921–27.

"Lead Poisoning in the United States," "Petroleum and Its Derivatives," "Coal Tar Benzene (Benzol) Poisoning," and "Phosphorus Poisoning: Sources, Symptoms and Treatment." In *Industrial Health,* edited by George M. Kober and Emery R. Hahurst. Philadelphia: P. Blakiston's Son and Company, 1924.

Industrial Poisons in the United States. New York: Macmillan, 1925.

"Fifteen Years in Industrial Toxicology." In *Contributions to Medical Science, Dedicated to Alfred Scott Warthin,* edited by Willard J. Stone. Ann Arbor, MI: G. Wahr, 1927.

"Recent Advances in Industrial Toxicology in the United States." *DeLamar Lectures.* The Johns Hopkins University School of Hygiene and Public Health. Baltimore: Williams and Williams, 1927.

Industrial Toxicology. Harper Medical Monographs. New York: Harper and Brothers, 1934.

"Because War Breeds War." In *Why Wars Must Cease,* edited by Rose Young. New York: Macmillan, 1935.

"Some New Developments in the Field of Volatile Solvents." In *Proceedings of the Occupational Disease Symposium.* Northwestern University Medical School, 1938.

Exploring the Dangerous Trades: The Autobiography of Alice Hamilton, M.D. Boston: Little, Brown and Company, 1943.

"From a Pioneer in the Poisonous Trades." In *Public Health in the World Today,* edited by James Stevens Simmons and Irene M. Kinsey. Cambridge, MA: Harvard University Press, 1949.

"Women in Harness." In *Five Years of Hitler,* edited by M. B. Schnapper. New York: American Council on Public Affairs, n.d.

With Harriet L. Hardy. *Industrial Toxicology.* 2nd rev. ed. New York: Paul B. Hoeber, 1949.

*With Harriet Hardy, *Industrial Toxicology*. 2nd and 3rd eds. Acton,
     MA: Publishing Science Group, 1949, 1974.
*Report of Commission on Occupational Diseases to His Excellency
     Governor Charles S. Daneen*. Chicago: Warner Printing, 1911.
International Congress of Women. *Report*. Amsterdam: N.V.
     Concordia, 1915.
With John F. Ransom. "Minority Report." *Report of the Health Insurance
     Commission of the State of Illinois*. Springfield: Illinois State Journal,
     1919.
Report of the Health Insurance Commission of the State of Illinois.
     Springfield: Illinois State Journal, 1919.

3. ARTICLES

"Do Women in Industry Need Special Health Legislation." *Consumer's
     League of Connecticut*, no. 12, n.d.
*"Ueber Einen Aus China Stammendon Kapsel Bacillus (Bacillus
     Capsulatus Chinesis Nov. Spec.)." *Zentral Watt Für Bakteriologie*,
     2 abt. 4 (1898): 230–36. As recorded in Angela Nugent Young,
     "Interpreting the Dangerous Trades: Workers Health in America
     and the Career of Alice Hamilton, 1910–1935" (PhD dissertation,
     Brown University, 1982), 212.
"Peculiar Form of Fibrosarcoma of the Brain." *The Journal of
     Experimental Medicine* 4 (September–November 1899):
     597–608.
"On the Presence of New Elastic Fibers in Tumors." *Journal of
     Experimental Medicine* 5, no. 2 (October 25, 1900): 131–38.
"The Pathology of a Case of Policencephalomyelitis." *Journal of Medical
     Research* (Boston) 8, no. 1 (June 1902): 11–30.
"The Fly as a Carrier of Typhoid: An Inquiry Into the Part Played by
     the Common House Fly in the Recent Epidemic of Typhoid Fever
     in Chicago." *Journal of the American Medical Association* 40, no. 9
     (February 28, 1903): 576–83.
"The Toxic Action of Scarlatinal and Pneumonic Sera on Paranoecia,"
     Journal of Infectious Diseases 1, no. 2 (March 19, 1904): 211–28.
"Surgical Scarlatina." *American Journal of the Medical Sciences*
     (Philadelphia) 128, no. 1 (July 1904): 111–28.

"The Question of Virulence Among the So-called Pseudo-Diphtheria
Bacilli." *Journal of Infectious Diseases* 1, no. 4 (November 5, 1904):
690–713.

"Dissemination of Streptococci Through Invisible Sputum." *Journal of
the American Medical Association* 44 (April 8, 1905): 1108–11.

"Milk and Scarlatina." *American Journal of the Medical Sciences*
(Philadelphia) 130, no. 5 (November 1905): 879–90.

"Occupational Conditions of Tuberculosis." In "The Industrial
Viewpoint," *Charities and the Commons* 16, no. 5 (May 5, 1906): 205–7.

"The Role of the House Fly and Other Insects in the Spread of Infectious
Diseases." *Illinois Medical Journal* 9, no. 6 (June 1906): 583–87.

"The Social Settlement and Public Health." *Charities and the Commons*
17 (March 9, 1907): 313–25.

"The Opsonic Index and Vaccine Therapy of Pseudodiphtheric Otitis."
Journal of Infectious Diseases 4, no. 3 (June 15, 1907): 313–25.

"Pseudodiphtheric Bacilli as Cause of Suppurative Otitis, Especially
Postcarlatinal." *Journal of Infectious Diseases* 4, no. 3 (June 15, 1907):
326–32.

"Gonorrheal Volvuvaginitis in Children." *Journal of Infectious Diseases* 5,
no. 2 (March 30, 1908): 133–57.

"Industrial Diseases With Special Reference to the Trades in Which
Women are Employed." *Charities and the Common* 20 (September 5,
1908): 655–59.

"Pathology and Bacteriology." *Charities and The Commons* 21
(November 7, 1908): 186–89.

"On the Occurence of Tharmostable and Simple Bactericidal and
Opsonic Substances," *Journal of Infectious Diseases* 5, no. 5
(December 18, 1908): 570–84.

"The Sociological Aspects of Medical Charities." *Chicago Medical
Recorder* (July 1909): 509–12.

"The Opsonic Index of Bacillus-Carriers." *Journal of the American
Medical Association* 54 (February 26, 1910): 704–5.

"Excessive Child Bearing as a Factor in Infant Mortality." *Bulletin of the
American Academy of Medicine* (April 1910).

"The Value of Opsonin Determinations in the Discovery of Typhoid
Carriers." *Journal of Infectious Diseases* 7, no. 3 (May 20, 1910):
393–410.

"Chronic Overwork." *Illinois Medical Journal* 17, no. 6 (June 1910): 739–43.

"Occupational Diseases." *Human Engineering* 1 (1911): 142–49.

"Lead Poisoning in Illinois." *American Labor Legislation Review* 1 (January 1911): 17–26.

"Lead Poisoning in Illinois." *Bulletin of the American Economic Association*, 4th Series, I (April 1911): 257–64.

"Lead Poisoning in Illinois." *Journal of the American Medical Association* 46 (April 29, 1911): 1240–44).

"What One Stockholder Did." *The Survey* 28 (June 1, 1912): 387–89.

"Industrial Lead Poisoning in Light of Recent Studies." *Journal of the American Medical Association* 49, no. 10 (September 7, 1912): 777–82.

"Industrial Plumbism." *Journal of the American Medical Association* 49, no. 14 (October 5, 1912): 1316.

"Fatigue: Smoke: Motherhood: And Other Equally Varied Factors Which Turn the World's Work into a Problem of Life and Health." *The Survey* 29, no. 5 (November 2, 1912): 152–54.

"Heredity and Responsibility." *The Survey* 29 (March 22, 1913): 865–66.

"The Friedman Cure." *The Survey* 30, no. 25 (September 20, 1913): 727–28.

"Unpaid Medical Service." *The Survey* 30, no. 25 (September 20, 1913): 727–28.

"Tuberculosis and the Hookworm in the Cotton Industry." *The Survey* 30 (September 20, 1913): 734–35.

"Leadless Glaze: What It Means to Pottery and Tile Workers." *The Survey* 31, no. 1 (October 4, 1913): 22–26.

"Rehabilitation of Boiled Milk." *The Survey* 31, no. 11 (December 13, 1913): 303.

"Radium Treatment of Cancer." *The Survey* 31 (January 31, 1914): 533–34.

"Lead-Poisoning in the United States." *American Journal of Public Health* 4 (June 1914): 477–80.

"War Surgery of Yesterday." *The Survey* 32 (September 5, 1914): 564–65.

"The Economic Importance of Lead Poisoning." *Bulletin of the American Academy of Medicine* 15 (October 1914): 299–304.

"Dammerschlaf." *The Survey* 33 (November 7, 1914): 158–59.

"At the War Capitals." *The Survey* 34 (August 7, 1915): 417–22, 433–36.

"Occupational Disease Clinic of New York City Health Department."
Monthly Review of the U.S. Bureau of Labor Statistics 1, no. 5
(November 1915): 7–19.

"What We Know About Cancer." *The Survey* 35 (November 20, 1915):
188–89.

"The Bollinger Case." *The Survey* 35 (December 4, 1915): 265–66.

"Race Suicide." *The Survey* 35 (January 1, 1916): 407–8.

"Is Science For or Against Human Welfare." *The Survey* 35 (February 5,
1916): 560–61.

"The Attitude of Social Workers Toward the War." *The Survey* 36 (June
17, 1916): 307–8.

"Wartime Economy and Hours of Labor." *The Survey* 36 (September 30,
1916): 638–30.

"As One Woman Sees the Issues." *The New Republic* 8 (October 7, 1916):
239–41.

"Health and Labor." *The Survey* 37 (November 11, 1916): 135–37.

"Industrial Poisons Used in the Making of Explosives." *Monthly Review
of the U.S. Bureau of Labor Statistics* 4, no. 2 (February 1917): 177–98.

"Prostitutes and Tuberculosis." *The Survey* 37, no. 18 (February 3, 1917):
516–17.

"Industrial Poisons Encountered in the Manufacture of Explosives."
Journal of the American Medical Association 58 (May 19, 1917): 1445–51.

"Toxic Jaundice in Munition Workers: A Review." *Monthly Review of the
U.S. Bureau of Labor Statistics* 5, no. 2 (August 1917): 63–74.

"Trinitrotoluene Poisoning." *Medicine and Surgery* 1 (September 1917): 761.

"Dope Poisoning in the Manufacture of Airplane Wings." *Monthly
Review of the U.S. Bureau of Labor Statistics* 5, no. 4 (October 1917):
18–25.

"Industrial Poisoning in Aircraft Manufacture." *Journal of the American
Medical Association* 69 (December 15, 1917): 2037–39.

"Dope Poisoning in the Making of Airplanes." *Monthly Review of the
U.S. Bureau of Labor Statistics* 6, no. 2 (February 1918): 37–64.

"Prophylaxis of Industrial Poisoning in the Munition Industries."
American Journal of Public Health 8 (February 1918): 125–30.

"The Fight Against Industrial Diseases: The Opportunities and Duties
of the Industrial Physician." *Pennsylvania Medical Journal* 31, no. 6
(March 1918): 378–80.

"Effect of the Air Hammer on the Hands of Stonecutters." *Monthly Review of the U.S. Bureau of Labor Statistics* 6 (May 1918): 237–50.

"Causation and Prevention of Trinitrotoluene (TNT) Poisoning." *Monthly Review of the U.S. Bureau of Labor Statistics* 6 (May 1918): 237–50.

"Industrial Poisoning in Aircraft Workers." *Bulletin of the International Association of Medical Museums*, no. 7 (May 1918): 97.

"Industrial Poisoning in Munition Workers." *Bulletin of the International Association of Medical Museums*, no. 7 (May 1918): 86.

"Dinitrophenol Poisoning in Munition Works in France." *Monthly Labor Review* 3, no. 3 (September 1918): 242–50.

"Practical Points in the Prevention of TNT Poisoning." *Monthly Labor Review* 8 (January 1919): 248–72.

"Industrial Poisoning in American Anilin Dye Manufacture." *Monthly Labor Review* 3, no. 2 (February 1919): 199–215.

"New Scientific Standards for Protection of Workers." *Proceedings of the Academy of Political Science* 8 (February 1919): 157–62.

"Lead Poisoning in American Industry." *Journal of Industrial Hygiene* 1, no. 1 (May 1919): 8–21.

"Prevention of TNT Poisoning." *American Journal of Public Health* 9 (May 1919): 394.

"Inorganic Poisons, Other than Lead, in American Industries." *Journal of Industrial Hygiene* 1, no. 2 (June 1919): 89–102.

"Medical and Surgical Lessons of the War: War Industrial Diseases." *Medical Record* 95 (June 21, 1919): 1053–59.

"Angels of Victory." *New Republic* 19 (June 25, 1919): 244–45.

"Health Hazards in the Manufacture of Dyestuffs." *Pennsylvania Medical Journal* 22 (July 1919): 170–80.

"Occupational Diseases in Pennsylvania." *Monthly Labor Review* 9, no. 1 (July 1919): 170–80.

"A Visit to Germany." *British Journal of Childhood Diseases* 16 (July–September 1919): 129–39.

"Industrial Poisoning by Compounds of the Aromatic Series." *Journal of Industrial Hygiene* 1, no. 4 (August 1919): 200–212.

"On a German Railway Train." *New Republic* 20 (September 24, 1919): 232–33.

"Hygienic Control of the Anilin Dye Industry in Europe." *Monthly Labor Review* 9 (December 1919): 1–21.

"A Discussion of the Etiology of So-called Aniline Tumors of the Bladder." *Journal of Industrial Hygiene* 3 (May 1921): 16–28.

"Trinitrotoluene as an Industrial Poison." *Journal of Industrial Hygiene* 3 (July 1921): 102–16.

"The Growing Menace of Benzene (Benzol) Poisoning in American Industry." *Journal of the American Medical Association* 78 (March 4, 1922): 627–30.

"Hazards in American Potteries." *The New Republic* 31 (July 12, 1922): 187.

"The Industrial Hygiene of Fur Cutting and Felt Hat Manufacture." *Journal of Industrial Hygiene* 4 (September 1922): 219–34.

"The Scope of the Problem of Industrial Hygiene." *Public Health Reports* 37, no. 42 (October 20, 1922): 2604–8.

"Sun-Baths for Rickets." *The Survey* 49 (November 15, 1922): 261.

"A Medieval Industry in the Twentieth Century." *The Survey* 51 (February 1, 1924): 456–64.

"Mercurialism in Quicksilver Production in California." *Journal of Industrial Hygiene* 5 (March 1924): 339–407.

"Protection for Working Women." *The Woman Citizen* 8 (March 8, 1924): 16–17.

"Protection of Women Workers." In "The 'Blanket' Amendment—A Debate." *The Forum* 72 (August 1924): 152–60.

"The Prevalence and Distribution of Industrial Lead Poisoning." *The Journal of the American Medical Association* 83 (August 23, 1924): 583–88.

"Doctor's Word on War." *The Woman Citizen* 9 (February 21, 1925): 15.

"What Price Safety? Tetra-Ethyl Lead Reveals a Flaw in Our Defenses." *The Survey* 54 (June 15, 1925): 333–34.

"Concerning Motor Car Gasoline." *The Woman Citizen* 10 (July 11, 1925): 14.

"Colonel House and Jane Addams." *The New Republic* 47 (May 26, 1926): 9–11.

"Witchcraft in West Polk Street." *The American Mercury* 10 (January 1927): 71–75.

"The Storage Battery Industry." *Journal of Industrial Hygiene* 9 (August 1927): 346–69.

"Recent Advances in Industrial Hygiene in Russia." *Rehabilitation Review* 1, no. 12 (December 1927).

"Eight Hour Day for Women in Industry." *Mid-Pacific Magazine* 36 (1928): 333–37.

"The Lessening Menace of Benzol Poisoning in American Industry."
    Journal of Industrial Hygiene 10 (September 1928): 227–33.
"Protection Against Industrial Poisoning." In *Chemistry in America*, edited
    by Julius Stieglitz (New York: The Chemical Foundation, 1929), as
    recorded in Young, "Interpreting the Dangerous Trades," 212.
"Poverty and Birth Control." New York League of Women Voters,
    February 1929.
"Enameled Sanitary Ware Manufacture." *Journal of Industrial Hygiene*
    11, no. 5 (May 1929): 139–53.
"The Cost of Medical Care." *The New Republic* 49 (June 26, 1929): 154.
"Nineteen Years in the Poisonous Trades." *Harper's Magazine* 159
    (October 1929): 580–91.
"A Vasomotor Disturbance in the Fingers of Stonecutters." *Archiv Fur
    Gewerbepathologie und Gewerbehygiene*, Band. I, Helft. 3 (January
    11, 1930): 348–58.
"State Pensions or Charity?" *Atlantic Monthly* 145 (May 1930): 683–87.
"Benzene (Benzol) Poisoning." *Archives of Pathology* 11 (March 1931):
    434–54; 11 (April 1931): 601–37.
"Methanol Poisoning." *New Republic* 66 (April 22, 1931): 277.
"Benzene Poisoning in Industry." *Medical Woman's Journal* 38
    (September 1931): 221–24.
"What about the Lawyers?" *Harper's Magazine* 163 (October 1931): 542ff.
"The Physical Risks in Child Labor." *New York Times*, May 22, 1932, III, 7.
"American and Foreign Labor Legislation: A Comparison." *Social Forces*
    11 (October 1932): 113–19.
"Industrial Hygiene." *American Journal of Public Health* 23 (April 1933):
    539–41.
"Formation of Phosgene in Thermal Decomposition of Carbon
    Tetrachloride." *Industrial and Engineering Chemistry* 25 (May 1933):
    539–41.
"Forward with Hitler." *Living Age* 344 (August 1933): 484.
"An Inquiry into the Nazi Mind." *New York Times*, August 6, 1933, VI,
    1–2.
"Industrial Poisons." *New England Journal of Medicine* 209 (August 10,
    1933): 279–81.
"Below the Surface." *Survey Graphic* 22 (September 1933): 449–54, 486.
"Hitler Speaks: His Book Reveals the Man." *Atlantic Monthly* 152, no. 4
    (October 1933): 399–408.

"The Youth Who Are Hitler's Strength." *New York Times*, October 8, 1933, VI, 3, 16.

"Sound and Fury in Germany." *Survey Graphic* 22 (November 1933): 549–54, 576, 578–79.

"The Plight of the German Intellectuals." *Harper's Magazine* 168 (January 1934): 159–69.

"Woman's Place in Germany." *Survey Graphic* 23 (January 1934): 26–29, 46–47.

"Industrial Poisons." *American Federationist* 43 (July 1936): 707–13.

"Some New and Unfamiliar Industrial Poisons." *New England Journal of Medicine* 215, no. 10 (September 3, 1936): 425–30.

"Medico-Legal Aspects of Industrial Poisoning." *Bulletin of the New York Academy of Medicine* 12 (December 1936): 637–49.

"Letters of a Woman Citizen." *Survey Graphic* 27 (May 1938): 282–83.

"Healthy, Wealthy—if Wise—Industry." *The American Scholar* 7, no. 1 (Winter 1938): 12–23.

"A Mid-American Tragedy." *Survey Graphic* 29 (August 1940): 434–37.

"Feed the Hungry." In "Shall We Feed Hitler's Victims?" *The Nation* 151 (December 14, 1940): 496–97.

"Exploring the Dangerous Trades." *Atlantic Monthly* 170 (December 1942): 109–25; 171 (January 1943): 117–32; 171 (February 1943): 117–32; 171 (March 1943): 131–46.

"The International Control of Disease After the War." *Connecticut Medical Journal* 7 (June 1943): 383–87.

"Toxicology of Chlorinated Hydrocarbons." *Yale Journal of Biology and Medicine* 15 (June 1943): 787–801.

"Death in the Factories." *The Nation* 157 (July 17, 1943): 66–68.

"Science in the Soviet Union." In "Industrial Medicine." *Science and Society* 8, no. 1 (1944): 69–73.

"New Problems in the Field of Industrial Toxicology." *California and Western Medicine* 61 (August 1944): 55–60.

"Diagnosis of Industrial Poisoning." *California and Western Medicine* 62 (March 1945): 110–12.

"Why I Am Against the Equal Rights Amendment." *Ladies Home Journal* 62 (July 1945): 23, 123.

"Looking at Industrial Nursing." *Public Health Nursing* 38 (February 1946): 63–65.

"To the Health of the Worker." *Nation's Business* 35 (May 1947): 56.

"Pioneering in Industrial Medicine." *Journal of the American Medical Women's Association* 2, no. 6 (June 1947).

"Forty Years in the Poisonous Trades." *American Industrial Hygiene Quarterly* 9, no. 1 (March 1948): 5–17.

"Industry is Health Conscious." *Medical Woman's Journal* 55, no. 10 (October 1948): 33–35, 64.

"The Encroachment of the State Upon the Conscience." *Christianity and Crisis* 11, no. 8 (1951): 61–62.

"Developments in the Field of Industrial Medicine." *Journal of the American Medical Women's Association* 6, no. 8 (August 1951): 313–14.

"Word's Lost, Strayed or Stolen." *Atlantic Monthly* 194 (September 1954): 55–56.

"English is a Queer Language." *Atlantic Monthly* 203 (June 1959): 51–52.

"A Woman of Ninety Looks at Her World." *Atlantic Monthly* 208 (September 1961): 51–55.

"Edith and Alice Hamilton: Students in Germany." *The Atlantic* 215 (March 1965): 129–32.

4. CO-AUTHORED ARTICLES

With J. M. Horton. "Further Studies on Virulent Pseduodiptheria Bacilli." *Journal of Infectious Diseases* 3, no. 1 (March 2, 1906): 128–47.

With Jean M. Cooke, "Inoculation Treatment of Gonorrheal Vulvovaginitis in Children." *Journal of Infectious Diseases* 5, no. 2 (March 30, 1908): 158–72.

With Rudolph W. Holmes, Caroline Hedger, Charles S. Bacon, and Herbert M. Stowe. "The Midwives of Chicago." *Journal of the American Medical Association* 50, no. 7 (April 25, 1908): 1346–50.

With Rey Vincent Luce. "Industrial Anilin Poisoning in the United States." *Journal of the American Medical Association* 66 (May 6, 1916): 1441–45.

"Industrial Anilin Poisoning in the United States." *Monthly Review of the U.S. Bureau of Labor Statistics* 2, no. 6 (June 1916): 1–12.

With Gertrude Seymour. "The New Public Health." *The Survey* 37
    (November 18, 1916): 166–69; 37 (January 20, 1917): 456–59; 38
    (April 21, 1917): 59–62.
With Jane Addams. "After the Lean years: Impressions of Food
    Conditions in Germany When Peace Was Signed." *The Survey* 42,
    no. 23 (September 6, 1919): 793–97.
With George Minot. "Ether Poisoning in the Manufacture of Smokeless
    Powder." *Journal of Industrial Hygiene* 3 (June 1920): 41–49.
With Rebecca Edith Hilles, "Industrial Hygiene in Moscow." *Journal of
    Industrial Hygiene* 7, no. 2 (February 1925): 41–61.
With Paul Reznikoff and Grace M. Burnham. "Tetra-Ethyl Lead."
    Journal of the American Medical Association 84 (May 16, 1925):
    1481–86.
With Jean M. Cooke, "Volatile Solvents Used in Industry." *American
    Journal of Public Health* 19 (May 1929): 523–26.
With Manfred Bowditch, C. K. Drinker, Philip Drinker, and H. H.
    Hagard, "A Code for Safe Concentrations of Certain Common
    Toxic Substances Used in Industry." *Journal of Industrial Hygiene
    and Toxicology* 22 (June 1940): 251.

MATTHEW C. RINGENBERG is Chair and
Professor of Social Work at Valparaiso University.

WILLIAM C. RINGENBERG is partially retired as
Professor Emeritus of History at Taylor University.

JOSEPH D. BRAIN is the Cecil K. and Philip Drinker
Professor of Environmental Physiology in the
Harvard University T.H. Chan School of Public Health.